THE WORLD BOOK ENCYCLOPEDIA OF
PEOPLE AND PLACES

5
O-S

WORLD
BOOK

a Scott Fetzer company
Chicago

www.worldbookonline.com

For information about other World Book publications,
visit our website at http://www.worldbookonline.com
or call 1-800-WORLDBK (1-800-967-5325).

For information about sales to schools and libraries, call
1-800-975-3250 (United States);
1-800-837-5365 (Canada).

Library of Congress Cataloging-in-Publication Data

The World Book encyclopedia of people and places.
    v. cm.
    Summary: "A 7-volume illustrated, alphabetically arranged
set that presents profiles of individual nations and other
political/geographical units, including an overview of history,
geography, economy, people, culture, and government of each.
Includes a history of the settlement of each world region
based on archaeological findings; a cumulative index; and Web
resources"--Provided by publisher.
    Includes index.
    ISBN 978-0-7166-3758-5
    1. Encyclopedias and dictionaries. 2. Geography--
Encyclopedias. I. World Book, Inc. Title: Encyclopedia of
people and places.
    AE5.W563 2011
    030--dc22
                                                      2010011919
This edition ISBN: 978-0-7166-3760-8

Printed in Hong Kong by Toppan Printing Co. (H.K.) LTD
3rd printing, revised, August 2012

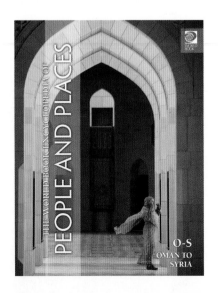

THE WORLD BOOK ENCYCLOPEDIA OF
PEOPLE AND PLACES

O-S
OMAN TO
SYRIA

Cover image:
Sultan Qaboos Grand
Mosque, Oman

© Pascal Deloche,
Godong/Corbis

CONTENTS

The world has 196 independent countries and about 50 dependencies. An independent country controls its own affairs. Dependencies are controlled in some way by independent countries. In most cases, an independent country is responsible for the dependency's foreign relations and defense, and some of the dependency's local affairs. However, many dependencies have complete control of their local affairs.

By 2010, the world's population was nearly 7 billion. Almost all of the world's people live in independent countries. Only about 13 million people live in dependencies.

Some regions of the world, including Antarctica and certain desert areas, have no permanent population. The most densely populated regions of the world are in Europe and in southern and eastern Asia. The world's largest country in terms of population is China, which has more than 1.3 billion people. The independent country with the smallest population is Vatican City, with only about 830 people. Vatican City, covering only 1/6 square mile (0.4 square kilometer), is also the smallest in terms of size. The world's largest nation in terms of area is Russia, which covers 6,601,669 square miles (17,098,242 square kilometers).

Every nation depends on other nations in some way. The interdependence of the entire world and its peoples is called *globalism*. Nations trade with one another to earn money and to obtain manufactured goods or the natural resources that they lack. Nations with similar interests and political beliefs may pledge to support one another in case of war. Developed countries provide developing nations with financial aid and technical assistance. Such aid strengthens trade as well as defense ties.

Nations of the World

Name	Map key		Name	Map key		Name	Map key
Afghanistan	D	13	Bulgaria	C	11	Dominican Republic	E	6
Albania	C	11	Burkina Faso	E	9	East Timor	F	16
Algeria	D	10	Burundi	F	11	Ecuador	F	6
Andorra	C	10‡	Cambodia	E	15	Egypt	D	1
Angola	F	10	Cameroon	E	10	El Salvador	E	5
Antigua and Barbuda	E	6	Canada	C	4	Equatorial Guinea	E	10
Argentina	G	6	Cape Verde	E	8	Eritrea	E	12
Armenia	D	12	Central African Republic	E	10	Estonia	C	1
Australia	G	16	Chad	E	10	Ethiopia	E	1
Austria	C	10	Chile	G	6	Federated States of Micronesia	E	1
Azerbaijan	D	12	China	D	14	Fiji	F	
Bahamas	D	6	Colombia	E	6	Finland	B	1
Bahrain	D	12	Comoros	F	12	France	C	10
Bangladesh	D	14	Congo, Democratic Republic of the	F	11	Gabon	F	10
Barbados	E	7	Congo, Republic of the	F	10	Gambia	E	
Belarus	C	11	Costa Rica	E	5	Georgia	C	12
Belgium	C	10	Côte d'Ivoire	E	9	Germany	C	10
Belize	E	5	Croatia	C	10	Ghana	E	
Benin	E	10	Cuba	D	5	Greece	D	1
Bhutan	D	14	Cyprus	D	11	Grenada	E	6
Bolivia	F	6	Czech Republic	C	10	Guatemala	E	
Bosnia-Herzegovina	C	10	Denmark	C	10	Guinea	E	
Botswana	G	11	Djibouti	E	12	Guinea-Bissau	E	
Brazil	F	7	Dominica	E	6			
Brunei	E	15						

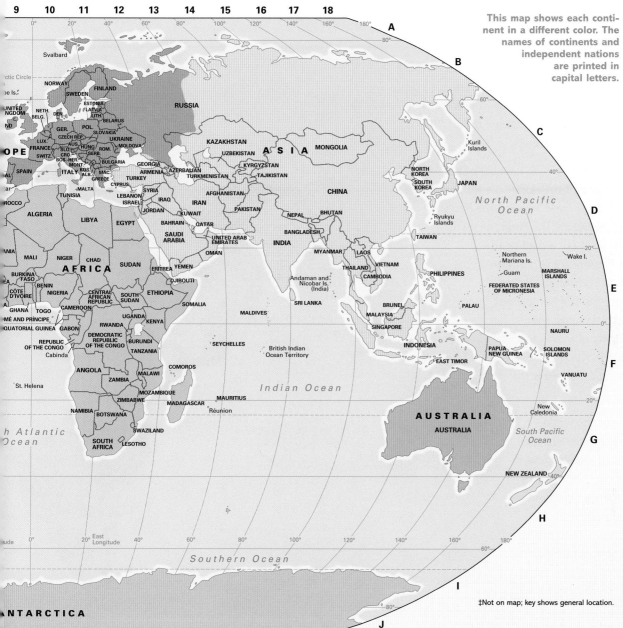

This map shows each continent in a different color. The names of continents and independent nations are printed in capital letters.

‡Not on map; key shows general location.

Name	Map key		Name	Map key		Name	Map key		Name	Map key		Name	Map key	
Guyana	E	7	Lebanon	D	11	Namibia	G	10	St. Vincent and the Grenadines	E	6	Taiwan	D	16
Haiti	E	6	Lesotho	G	11	Nauru	F	18	Samoa	F	1	Tajikistan	D	14
Honduras	E	5	Liberia	E	9	Nepal	D	14	San Marino	C	10‡	Tanzania	F	11
Hungary	C	10	Libya	D	10	Netherlands	C	10	São Tomé and Príncipe	E	10	Thailand	E	15
Iceland	B	9	Liechtenstein	C	10‡	New Zealand	G	18	Saudi Arabia	D	12	Togo	E	9
India	D	13	Lithuania	C	11	Nicaragua	E	5	Senegal	E	9	Tonga	F	1
Indonesia	F	16	Luxembourg	C	10	Niger	E	10	Serbia	C	10	Trinidad and Tobago	E	6
Iran	D	12	Macedonia	C	11	Nigeria	E	10	Seychelles	F	12	Tunisia	D	10
Iraq	D	12	Madagascar	F	12	Norway	B	10	Sierra Leone	E	9	Turkey	D	11
Ireland	C	9	Malawi	F	11	Oman	E	12	Singapore	E	15	Turkmenistan	D	13
Israel	D	11	Malaysia	E	15	Pakistan	D	13	Slovakia	C	11	Tuvalu	F	1
Italy	C	10	Maldives	E	13	Palau	E	16	Slovenia	C	11	Uganda	E	11
Jamaica	E	6	Mali	E	9	Panama	E	5	Solomon Islands	F	18	Ukraine	C	11
Japan	D	16	Malta	D	10	Papua New Guinea	F	17	Somalia	E	12	United Arab Emirates	D	12
Jordan	D	11	Marshall Islands	E	18	Paraguay	G	7	South Africa	G	11	United Kingdom	C	9
Kazakhstan	C	13	Mauritania	D	9	Peru	F	6	Spain	C	9	United States	C	4
Kenya	E	11	Mauritius	G	12	Philippines	E	16	Sri Lanka	E	14	Uruguay	G	7
Kiribati	F	1	Mexico	D	4	Poland	C	10	Sudan	E	11	Uzbekistan	D	14
Korea, North	C	16	Moldova	C	11	Portugal	D	9	Sudan, South	E	11	Vanuatu	F	18
Korea, South	D	16	Monaco	C	10‡	Qatar	D	12	Suriname	E	7	Vatican City	C	10‡
Kosovo	C	11	Mongolia	C	15	Romania	C	11	Swaziland	G	11	Venezuela	E	6
Kuwait	D	12	Montenegro	C	10	Russia	C	13	Sweden	B	10	Vietnam	E	15
Kyrgyzstan	C	13	Morocco	D	9	Rwanda	F	11	Switzerland	C	10	Yemen	E	12
Laos	E	15	Mozambique	F	11	St. Kitts and Nevis	E	6	Syria	D	11	Zambia	F	11
Latvia	C	11	Myanmar	D	14	St. Lucia	E	6				Zimbabwe	G	11

PHYSICAL WORLD MAP

The surface area of the world totals about 196,900,000 square miles (510,000,000 square kilometers). Water covers about 139,700,000 square miles (362,000,000 square kilometers), or 71 percent of the world's surface. Only 29 percent of the world's surface consists of land, which covers about 57,200,000 square miles (148,000,000 square kilometers).

Oceans, lakes, and rivers make up most of the water that covers the surface of the world. The water surface consists chiefly of three large oceans—the Pacific, the Atlantic, and the Indian. The Pacific Ocean is the largest, covering about a third of the world's surface. The world's largest lake is the Caspian Sea, a body of salt water that lies between Asia and Europe east of the Caucasus Mountains. The world's largest body of fresh water is the Great Lakes in North America. The longest river in the world is the Nile in Africa.

The land area of the world consists of seven continents and many thousands of islands. Asia is the largest continent, followed by Africa, North America, South America, Antarctica, Europe, and Australia. Geographers sometimes refer to Europe and Asia as one continent called Eurasia.

The world's land surface includes mountains, plateaus, hills, valleys, and plains. Relatively few people live in mountainous areas or on high plateaus since they are generally too cold, rugged, or dry for comfortable living or for crop farming. The majority of the world's people live on plains or in hilly regions. Most plains and hilly regions have excellent soil and an abundant water supply. They are good regions for farming, manufacturing, and trade. Many areas unsuitable for farming have other valuable resources. Mountainous regions, for example, have plentiful minerals, and some desert areas, especially in the Middle East, have large deposits of petroleum.

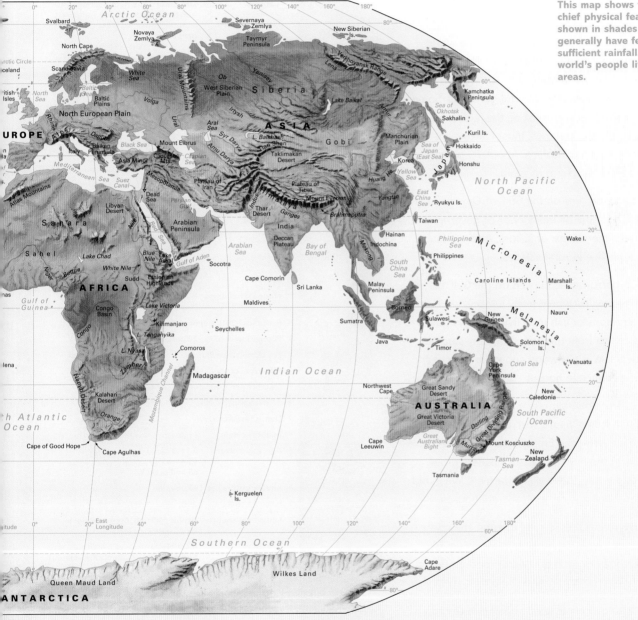

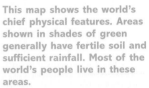

This map shows the world's chief physical features. Areas shown in shades of green generally have fertile soil and sufficient rainfall. Most of the world's people live in these areas.

Arctic Ocean

Svalbard

Novaya Zemlya

North Cape

Scandinavia

Iceland

Arctic Circle

British Isles

North Sea

Baltic Sea

Baltic Plains

North European Plain

EUROPE

Alps

Danube

Italy

Balkan Peninsula

Asia Minor

Mediterranean Sea

Suez Canal

Atlas Mountains

Sahara

Sahel

Libyan Desert

Lake Chad

Niger

Benue

White Nile

Blue Nile

Lake Assal

AFRICA

Sudd

Ethiopian Highlands

Congo Basin

Congo

Kilimanjaro

Lake Victoria

L. Tanganyika

L. Nyasa

Zambezi

Comoros

Madagascar

Mozambique Channel

Namib Desert

Kalahari Desert

Orange

Cape of Good Hope

Cape Agulhas

Gulf of Guinea

Helena

South Atlantic Ocean

Severnaya Zemlya

Taymyr Peninsula

New Siberian Is.

Ob

West Siberian Plain

Siberia

Ural Mountains

Volga

Ural

Irtysh

Yenisey

Lena

Verkhoyansk Range

Lake Baikal

Amur

ASIA

Gobi

Aral Sea

Caspian Sea

Black Sea

Mount Elbrus

Caucasus Mts.

Syr Darya

L. Balkhash

Tien Shan

Amu Darya

Taklimakan Desert

Mesopotamia

Plateau of Iran

Persian Gulf

Dead Sea

Red Sea

Nile

Arabian Peninsula

Gulf of Aden

Socotra

Plateau of Tibet

Himalaya

Mount Everest

Indus

Thar Desert

Ganges

Brahmaputra

Yangtze

Huang He

Manchurian Plain

Korea

Yellow Sea

East China Sea

Japan

Sea of Okhotsk

Kamchatka Peninsula

Sakhalin

Kuril Is.

Hokkaido

Honshu

Ryukyu Is.

Taiwan

Hainan

Indochina

South China Sea

Philippine Sea

Philippines

Micronesia

Wake I.

Caroline Islands

Marshall Is.

Nauru

Melanesia

New Guinea

Solomon Is.

Vanuatu

New Caledonia

India

Deccan Plateau

Arabian Sea

Cape Comorin

Bay of Bengal

Sri Lanka

Maldives

Seychelles

Sumatra

Java

Malay Peninsula

Borneo

Sulawesi

Timor

North Pacific Ocean

Indian Ocean

Coral Sea

Cape York Peninsula

Great Sandy Desert

Northwest Cape

AUSTRALIA

Great Victoria Desert

Great Australian Bight

Cape Leeuwin

Darling

Murray

Great Dividing Range

Mount Kosciuszko

New Zealand

Tasman Sea

Tasmania

South Pacific Ocean

Kerguelen Is.

Southern Ocean

Queen Maud Land

Wilkes Land

Cape Adare

ANTARCTICA

7

OMAN

Oman is a small country on the southeastern corner of the Arabian Peninsula. It includes the tip of the mountainous Musandam Peninsula to the north, which is separated from the rest of Oman by the United Arab Emirates. From its position near the mouth of the Persian Gulf, Oman watches much of the world's oil pass on its way to nations around the globe.

Oman is one of the hottest countries in the world. Most Omani men wear white robes and turbans to protect themselves from the blazing sun. The women wear long, black outer dresses over colorful garments. Some wear black masks that cover most of the face, a practice common in Islamic countries. Most Omanis are Arabs, and almost all are Muslims. About 75 percent follow Ibadi, a moderate form of Islam.

The capital, Muscat, lies on the Gulf of Oman. Many city people work for the oil industry or as government officials, laborers, merchants, or sailors. About a fourth of Oman's people live in rural villages.

Al Batinah, a narrow coastal plain on the Gulf of Oman, is a fertile region where many Omanis work on date palm plantations or fish for a living. Crews pull in large catches of fish from the gulf, especially sardines. Coastal villagers live in old wood and palm-thatched homes or in new concrete houses.

A rugged, steep mountain range called Al Hajar separates Al Batinah from Oman's vast, arid interior. The desert of Rub al Khali—the Empty Quarter—covers western Oman. In the interior, Omani villagers grow dates, fruits, and grain. These village farmers live in old mud and stone dwellings or new concrete houses.

Nomads also roam Oman's rural interior. They wander from place to place with their animals, living in tents and searching for food and water for their herds.

In a region called Dhofar in southwestern Oman, enough rain falls to allow tropical vegetation to grow. Farmers cultivate such fruits as bananas, coconuts, and limes, and many also raise cattle. Dhofar is famous for the frankincense trees that grow on a plateau just north of the Jabal al Qara, a mountain range that hugs the southwest coast.

FACTS

Official name:	Saltanat Uman (Sultanate of Oman)
Capital:	Muscat
Terrain:	Vast central desert plain, rugged mountains in north and south
Area:	119,499 mi^2 (309,500 km^2)
Climate:	Dry desert; hot, humid along coast; hot, dry interior; strong southwest summer monsoon (May to September) in far south
Main rivers:	N/A
Highest elevation:	Jabal Ash Sham, 9,957 ft (3,035 m)
Lowest elevation:	Arabian Sea, sea level
Form of government:	Monarchy
Head of state:	Sultan and prime minister
Head of government:	Sultan and prime minister
Administrative areas:	5 mintaqat (regions), 4 muhafazat (governorates)
Legislature:	Majlis Oman consisting of the Majlis ad-Dawla (upper chamber) with members appointed by the sultan and the Majlis ash-Shura (lower chamber) with 84 members serving four-year terms
Court system:	Supreme Court
Armed forces:	42,600 troops
National holiday:	Birthday of Sultan Qaboos - November 18 (1940)
Estimated 2010 population:	2,815,000
Population density:	24 persons per mi^2 (9 per km^2)
Population distribution:	72% urban, 28% rural
Life expectancy in years:	Male, 72; female, 76
Doctors per 1,000 people:	1.7
Birth rate per 1,000:	24
Death rate per 1,000:	3
Infant mortality:	11 deaths per 1,000 live births
Age structure:	0-14: 32%; 15-64: 65%; 65 and over: 3%
Internet users per 100 people:	17
Internet code:	.om
Languages spoken:	Arabic (official), English, Balochi, Urdu, Indian dialects
Religions:	Ibadi Muslim 75%, other (including Sunni Muslim, Shiah Muslim, Hindu) 25%
Currency:	Omani rial
Gross domestic product (GDP) in 2008:	$56.32 billion U.S.
Real annual growth rate (2008):	6.7%
GDP per capita (2008):	$20,821 U.S.
Goods exported:	Crude oil and refined petroleum, fish products, food, natural gas
Goods imported:	Food, iron and steel, machinery, motor vehicles
Trading partners:	China, India, Japan, United Arab Emirates, United States

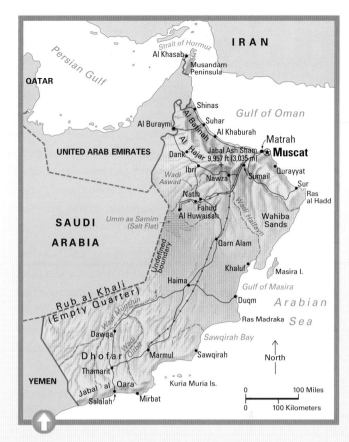

An Omani woman displays traditional Omani dress—a colorful inner robe covered with a long, black outer dress. In this strict Islamic society, many women also wear black veils.

Oman lies in an important position at the outlet of the Persian Gulf. The Gulf of Oman and the Arabian Sea border eastern Oman, providing the country with access to the ocean. Oman's northern tip lies on the strategic Strait of Hormuz—through which much of the world's oil is shipped.

Muscat is Oman's capital and largest city. Muscat lies on the country's northeast coast, along the Gulf of Oman. The port-district of Muttrah forms the northeastern edge of the city. A blue-tiled minaret towers above the surrounding buildings.

Today, oil exports account for most of Oman's income. But Oman was an extremely poor country, with an economy based on farming and fishing, until 1964, when oil was discovered. Most Omanis still struggle to make a living, but oil money has been used to finance improvements.

The economy began to change in 1970, when Sultan Said bin Taimur, who was opposed to modernization, was overthrown by his son, Qaboos bin Said. As the new sultan, Qaboos developed the oil industry. With the oil income, he built roads, hospitals, and schools.

Before Qaboos became sultan, few Omani children ever went to school. Qaboos encouraged Omanis to educate their children. He also helped farmers by promoting irrigation techniques and other modern methods. Omanis traditionally get their water from wells, some of which are fed by underground canals that were built in ancient times.

Sultan Qaboos is a member of the Al Bu Said family, which has ruled Oman since the 1740's. Since 1798, Oman and Great Britain (now the United Kingdom) have kept close ties. The British helped Qaboos's father put down a religious rebellion in 1959.

As ruler of Oman, the sultan appoints a cabinet to carry out the government's operations. Two councils advise the sultan. The sultan appoints the Council of State. The people elect the Consultative Council, whose members serve four-year terms.

THE PACIFIC ISLANDS

The Pacific Islands, or Oceania, is a group of about 20,000 to 30,000 islands scattered across the Pacific Ocean. Some islands cover thousands of square miles, while others are no more than tiny piles of rock or sand that barely rise above the water. Some Pacific Islands, particularly New Guinea and New Zealand, are also considered part of Australasia. The Pacific Islands are often divided into three main areas: (1) Melanesia, to the southwest; (2) Micronesia, to the northwest; and (3) Polynesia, to the east.

Not considered as among the Pacific Islands group are some islands near the mainlands of Asia, North America, and South America. For example, the islands that make up the nations of Indonesia, Japan, and the Philippines are considered part of Asia, and the Aleutians and the Galapagos are grouped with North America and South America, respectively. Australia, too, is not considered part of the group because it is a continent.

Together, the Pacific Islands have an area of under 490,000 square miles (1,270,000 square kilometers)—less than the state of Alaska. New Guinea and the two main islands of New Zealand make up about 90 percent of the total land area. About 18 million people live in the Pacific Islands. Only a few islands or island groups have large populations, while many have fewer than a hundred people. Many others have no people at all.

Melanesians inhabit the islands of the southwestern Pacific. The name *Melanesia* comes from Greek words meaning *black islands*. Early visitors to the region gave it this name because of the dark skin of its inhabitants. Melanesia today includes the independent countries of Papua New Guinea, the Solomon Islands, and Vanuatu, as well as the French territory of New Caledonia. The country of Fiji is also included in Melanesia, even though it shares many cultural characteristics with Polynesia. Most Melanesian islands lie south of the equator.

Micronesians inhabit the islands of the central and northwestern Pacific. The name *Micronesia* comes from Greek words meaning *tiny islands*. These islands lie north of Melanesia, and most of them also lie north of the equator. Micronesia includes Guam, the Commonwealth of the Northern Marianas, the Federated States of Micronesia, Nauru, Palau, and the Republics of Kiribati and of the Marshall Islands.

Polynesians are the peoples of the eastern Pacific. *Polynesia* comes from Greek words meaning *many islands*. Polynesia occupies the largest area in the South Pacific. American Samoa, the

Cook Islands, French Polynesia, Hawaii, New Zealand, Niue, Samoa, Tokelau, Tonga, Tuvalu, and Wallis and Futuna are in Polynesia.

The Pacific Islands are also home to many migrant groups. Today, Chinese communities can be found throughout much of Polynesia. Americans, Chinese, Europeans, Filipinos, and Japanese live in the islands of Micronesia. Much of Melanesia is home to people of Asian and European heritage. Large French populations exist in French Polynesia and New Caledonia, and most New Zealanders have British ancestors.

Land and climate

The land and climate vary greatly among the Pacific Islands. Many of the islands, especially those in Polynesia, are noted for their sparkling white beaches, gentle ocean breezes, and swaying palm trees. Other islands, especially in Melanesia, have thick jungles and tall mountain peaks. Many lowland areas in these islands are extremely hot and humid, but snow covers the tallest mountain peaks throughout the year.

Economy

The Pacific Ocean is of great importance to the people who live on the Pacific Islands. The ocean provides low-cost sea transportation, extensive fishing waters, offshore oil and gas fields, minerals, and sand and gravel for the construction industry. Much of the world's fish catch comes from Oceania. Offshore oil and gas reserves provide a vital supply of energy for Australia, China, New Zealand, Peru, and the United States. In addition, the United States has military bases in Guam, Hawaii, and the Marshall Islands that greatly influence the economies of those areas.

Fiji, Hawaii, and New Zealand have well-developed economies with a large number of agricultural and manufacturing industries. Mining is a major industry in New Caledonia, Papua, and Papua New Guinea. Tourism and government employment are also major sources of income in most of the Pacific. As growing numbers of tourists visit the region, the islands are working to build more airports, hotels, highways, shops, and restaurants. However, some islanders fear that further growth of the tourist industry will destroy the natural charm and traditional ways of the Pacific.

Most Pacific Islanders have access to land and to food they have grown themselves. They live in what economists call an *informal economy,* sometimes earning income from selling betel nut, *copra* (the dried meat of the coconut), fish, fruits, kava, local crafts, sweet potato, taro, and yams.

Most Pacific Island countries receive financial aid from other countries and from international organizations. Many Pacific Islanders also receive support from family members living overseas.

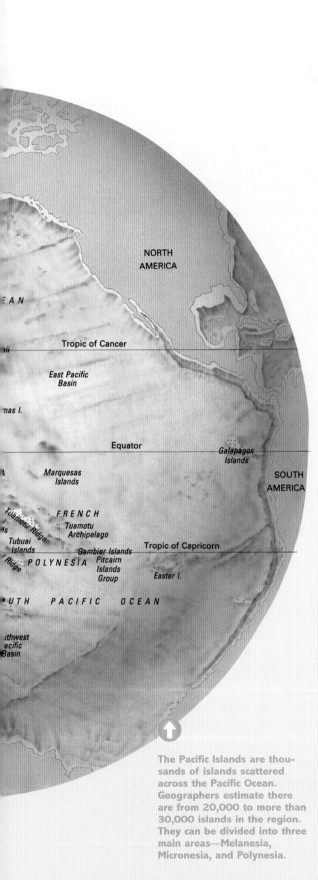

The Pacific Islands are thousands of islands scattered across the Pacific Ocean. Geographers estimate there are from 20,000 to more than 30,000 islands in the region. They can be divided into three main areas—Melanesia, Micronesia, and Polynesia.

HISTORY

The first settlers in the Pacific Islands probably came from Southeast Asia thousands of years ago. They sailed to the islands on rafts or dugout canoes and followed land bridges whenever possible. Over many centuries, people settled on the main islands of Melanesia, to the southwest, and Micronesia, to the northwest. The main islands of Polynesia, to the east, were settled later. With large expanses of ocean separating the islands, people in distant island groups had little or no contact.

Beginning in the early 1500's, European explorers began to visit the Pacific Islands. The most famous Pacific explorer of the 1700's was Captain James Cook of the British Royal Navy. His discoveries encouraged Protestants and Roman Catholics to establish missions throughout the area. Many island peoples who today are Christians are descendants of those converted during the 1800's. Many missionaries introduced genuine improvements to the islands, but others concentrated largely on doing away with native customs and traditions.

Traders searching for coconut oil, sandalwood, and other products came to the islands. Whaling vessels also stopped there. New settlers included many criminals and drifters, and lawlessness became a problem. Diseases brought by Europeans killed many islanders.

By the 1800's, France, Germany, Spain, the United Kingdom, and the United States were competing for control of the Pacific Islands, and each acquired several islands or island groups. After Spain's defeat in the Spanish-American War of 1898, Germany and the United States took over the Spanish possessions in Micronesia. By the early 1900's, Germany also held parts of Nauru, New Guinea, and Samoa. After Germany's defeat in World War I (1914–1918), control of its Pacific Islands was transferred to Japan, New Zealand, and Australia. Through all these changes of rule, the islanders had little or no voice in their government.

Japan increased its power in the Pacific after World War I. Finally, in December 1941, Japanese bombers attacked the U.S. naval base at Pearl Harbor, Hawaii, and opened the Pacific theater of World War II (1939–1945). By mid-1942, Japanese troops had captured islands as

Captain Cook landing at Tanna New Hebrides.

The catamaran, a raftlike boat with two hulls, carried the people of the Pacific Islands on voyages between island groups. In 1774, Captain Cook mapped the region in Melanesia that he named the New Hebrides. The islands are now called Vanuatu.

far east as the Gilberts and as far south as the Solomons. The United States then began to drive the Japanese off these islands. In September 1945, Japan surrendered, losing its huge Pacific empire.

After World War II, the United Nations decided that four areas in the Pacific should be governed as trust territories until they were ready for independence. New Zealand administered Western Samoa as a trust territory until 1962, when it gained independence. Western Samoa later changed its name to Samoa. Australia, the United Kingdom, and New Zealand governed Nauru as a trust territory until 1968, when it became independent. Australia governed the Trust Territory of New Guinea until 1973, when it became part of the self-governing territory of Papua New Guinea. Papua New Guinea became independent in 1975. The United States administered the Trust Territory of the Pacific Islands.

In 1986, all of the Mariana Islands except Guam became a commonwealth of the United States. Guam is a U.S. territory. Also in 1986, the Marshall Islands and the

Federated States of Micronesia became independent nations in *free association* with the United States. Under free association, their governments control their internal and foreign affairs, and the United States is obligated to defend the islands in emergency situations. In 1994, the Palau Islands became the independent nation of Palau, also in free association with the United States.

Many Pacific countries have struggled with conflicts between traditional cultural concepts and new ideas. Many people of the Pacific Islands still live in the same kinds of houses, eat the same kinds of food, and wear the same kinds of clothing that their ancestors did. But these traditional ways of life are changing rapidly, as an increasing number of people adopt Western customs and leave their villages to work in towns and cities.

The settlement of the Pacific Islands occurred thousands of years ago. The first inhabitants came from Asia and moved from island to island. Melanesia and Micronesia, two of the three main cultural areas of the Pacific Islands, were settled first. Polynesia, the third cultural area, was settled last. A Polynesian people, the Maori, arrived in New Zealand during the A.D. 900's. Europeans began exploring the Pacific Islands in the 1500's, following Magellan's voyage across the Pacific. By the late 1800's, France, Germany, Spain, the United Kingdom, and the United States were competing for control of the Pacific Islands. Japan increased its power in the Pacific after World War I. Since World War II, many island groups have gained their independence.

JAPAN

Midway (USA)

Wake Island (USA)

NORTH

PACIFIC

OCEAN

Mendaña (1567-69)

Hawaiian Islands (USA)

CHINA

MICRONESIA

NORTHERN MARIANA ISLANDS (USA)

Saipan
Tinian

MARSHALL ISLANDS (USA)

Bikini

Enewetak

GUAM (USA)

Kwajalein

Philippine Sea

FEDERATED STATES

Yap Truk Pohnpei
OF
Caroline Islands
MICRONESIA (USA)

Kosrae

PHILIPPINES

PALAU (USA Trust)

NAURU

Gilbert Islands (KIR)

Banaba (KIR)

KIRIBATI

Line Islands (KIR and USA)

Bismarck Archipelago (PNG)

PAPUA NEW GUINEA

SOLOMON ISLANDS

Bougainville

TUVALU (formerly Ellice Islands)

Phoenix Islands (KIR)

Tokelau Islands (NZ)

Marquesas Islands (FR)

MELANESIA

VANUATU (formerly New Hebrides)

FIJI

WALLIS AND FUTUNA (FR)

SAMOA

AMERICAN SAMOA (USA)

Society Islands (FR)

Tahiti (FR)

FRENCH

POLYNESIA (FR)

Tuamotu Archipelago

INDONESIA

Arafura Sea

TONGA

Niue (NZ)

Cook Islands (NZ)

Tasman (1642)

Coral Sea

Loyalty Islands (FR)

NEW CALEDONIA (FR)

POLYNESIA

Tubuai Islands (FR)

Magellan 1521

PITCAIRN ISLANDS GROUP (BR)

AUSTRALIA

Norfolk (AUST)

Kermadec Islands (NZ)

SOUTH PACIFIC

OCEAN

EASTER ISLAND (CHILE)

Tasman Sea

NEW ZEALAND

North Island

Cook (1768-71)

South Island

Cook (1776-79)

Pakistan is a Muslim nation in South Asia. It is bordered on the west by Iran and Afghanistan, on the north by China, and on the east by India. Its southern coast lies along the Arabian Sea. Pakistan is a land of soaring mountains, high plateaus, fertile plains, and hot, barren deserts. The Indus is Pakistan's major river.

The history of the region that is now Pakistan can be traced back about 4,500 years. However, it was not until 1947 that Pakistan became an independent nation. Since then, Pakistan has gone through many political and geographical changes. Numerous different cultural groups live in Pakistan, each with its own customs and language. These cultural differences have made it difficult for Pakistan to become a strong, unified nation.

Early history

The Indus River Valley in Pakistan is known as one of the cradles of civilization because of the important civilization that developed there about 2500 B.C. Ruins of that great culture's two cities, Harappa and Mohenjo-Daro, still survive today. These ruins show that it was a large, well-planned civilization. For unknown reasons, the Indus Valley civilization disappeared about 1700 B.C.

Throughout the following centuries, the Indus Valley was conquered by many different peoples. It was once part of the Achaemenid (Persian) Empire. Later, Alexander the Great conquered the region.

After the rise and fall of the Maurya Empire about 230 B.C., the Greeks ruled the Indus region. They were conquered by the Scythians from Afghanistan, who were replaced by the Parthians. The Kushan Empire succeeded the Parthians, and during the A.D. 300's, the region became part of the Gupta Empire. The Huns ruled the Indus in the mid-400's.

PAKISTAN

The coming of Islam

Around 712, Arab Muslims sailed across the Arabian Sea, bringing the religion of Islam to the region. About 300 years later, Turkish Muslims established a Muslim kingdom in the Indus Valley. Later, the Delhi Sultanate, a Muslim empire, took over until Babur, a Muslim ruler from Afghanistan, established the Mogul Empire in 1526.

Beginning in the 1500's, the British East India Company, which already had a firm trading foothold in India, became even stronger in the region. Eventually, the company gained political power over much of India. When the British government took over the country in 1857, the region came to be known as *British India*.

An independent nation

In 1947, the United Kingdom gave in to the demands of the Indian people for independence. The country was divided into two separate nations according to its major religious groups. India was designated the Hindu state. The new state of Pakistan was made up of the two territories inhabited by Muslims—East Pakistan and West Pakistan. These territories were more than 1,000 miles (1,600 kilometers) apart.

West Pakistan was made up of a mixture of the provinces of Sindh, Balochistan, the Khyber Pakhtunkhwa, parts of Punjab, and a number of small principalities. The people of East Pakistan were mainly Bengalis. Although the people of East Pakistan and West Pakistan shared the religion of Islam, important cultural differences, as well as a great distance, separated them. In 1971, civil war erupted between East and West Pakistan. On March 26, 1971, East Pakistan declared itself an independent nation called Bangladesh.

PAKISTAN TODAY

Since becoming an independent nation in 1947, Pakistan has been troubled by geographical disputes and political instability. Civil war has damaged its economy. The cultural differences between its many ethnic groups have made it difficult to unite the people in solving the country's problems.

Each of Pakistan's four provinces—Balochistan, Khyber Pakhtunkhwa, the Punjab, and Sindh—consists of a variety of cultural groups. The Punjab in the northeast has the largest population. As a result, the Punjabis have controlled the government, economy, and armed forces of Pakistan through much of the nation's history.

Civil war

On March 26, 1971, the civil war between East and West Pakistan intensified, when East Pakistan declared itself the independent nation of Bangladesh. India joined Bangladesh in its fight against West Pakistan in December. Two weeks later, Pakistan surrendered.

Agha Mohammad Yahya Khan, the president who had tried to keep East and West Pakistan together, resigned. Pakistan lost much of its area and more than half its people. Its economy was badly disrupted.

The late 1900's

Zulfikar Ali Bhutto succeeded Yahya Khan as president of Pakistan. Under Bhutto's leadership, Pakistan adopted a new constitution in 1973 that established Islamic socialism as the nation's guiding principle and retained the structure of a federal republic. Under the new constitution, Bhutto became prime minister and Chaudhri Fazal Elahi became president.

In 1977, when parliamentary elections resulted in a victory for Bhutto's party, the Pakistan People's Party (PPP), many people accused the party of election fraud. Violence broke out. Military officers led by General Mohammed Zia-ul-Haq removed Bhutto from office. Zia then suspended the constitution and declared martial law. In 1978, Zia declared himself president. Bhutto was executed in 1979.

FACTS

Official name:	Islamic Republic of Pakistan
Capital:	Islamabad
Terrain:	Flat Indus plain in east; mountains in north and northwest; Balochistan plateau in west
Area:	307,374 mi^2 (796,095 km^2)
Climate:	Mostly hot, dry desert; temperate in northwest; arctic in north
Main rivers:	Indus, Jhelum, Chenab
Highest elevation:	K2 (in Kashmir), 28,250 ft (8,611 m)
Lowest elevation:	Indian Ocean, sea level
Form of government:	Federal republic
Head of state:	President
Head of government:	Prime minister
Administrative areas:	4 provinces, 1 territory, 1 capital territory
Legislature:	Parliament consisting of the National Assembly with 342 members serving five-year terms and the Senate with 100 members serving six-year terms
Court system:	Supreme Court, Federal Shariat Court
Armed forces:	617,000 troops
National holiday:	Republic Day - March 23 (1956)
Estimated 2010 population:	173,117,000
Population density:	563 persons per mi^2 (217 per km^2)
Population distribution:	65% rural, 35% urban
Life expectancy in years:	Male, 63; female, 65
Doctors per 1,000 people:	0.8
Birth rate per 1,000:	28
Death rate per 1,000:	8
Infant mortality:	73 deaths per 1,000 live births
Age structure:	0-14: 37%; 15-64: 59%; 65 and over: 4%
Internet users per 100 people:	11
Internet code:	.pk
Languages spoken:	Punjabi, Sindhi, Siraiki, Pashtu Urdu (official), English (official)
Religions:	Sunni Muslim 77%, Shiah Muslim 20%, other (including Christian and Hindu) 3%
Currency:	Pakistani rupee
Gross domestic product (GDP) in 2008:	$165.61 billion U.S.
Real annual growth rate (2008):	5.8%
GDP per capita (2008):	$986 U.S.
Goods exported:	Carpets and rugs; leather goods; rice; textiles, including garments, cotton cloth, and yarn
Goods imported:	Chemicals, iron and steel, machinery, petroleum and petroleum products, transportation equipment
Trading partners:	China, Japan, Saudi Arabia, United Arab Emirates, United States

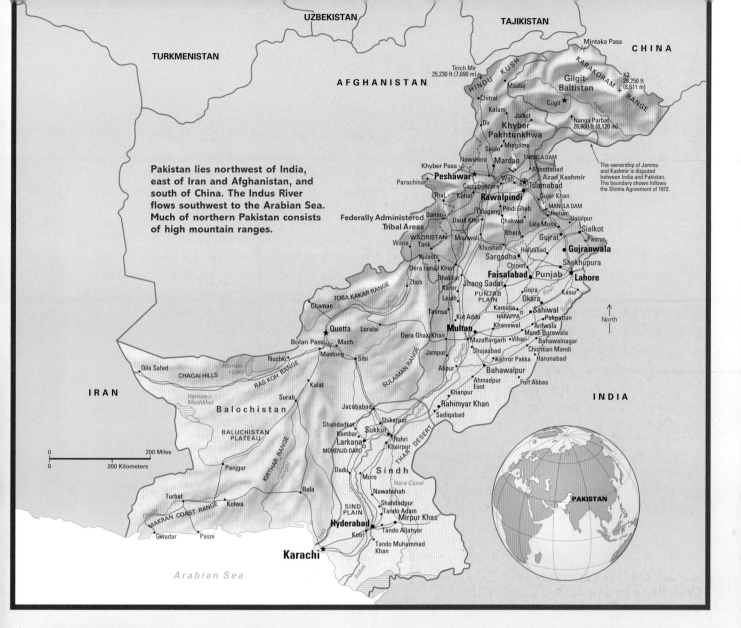

Pakistan lies northwest of India, east of Iran and Afghanistan, and south of China. The Indus River flows southwest to the Arabian Sea. Much of northern Pakistan consists of high mountain ranges.

The ownership of Jammu and Kashmir is disputed between India and Pakistan. The boundary shown follows the Shimla Agreement of 1972.

In August 1988, Zia died in a plane crash. Later that year, parliamentary elections were held. Benazir Bhutto, head of the PPP and daughter of Zulfikar Ali Bhutto, became prime minister and head of the government. In 1990, Pakistan's president dismissed Bhutto's government. Nawaz Sharif of the Pakistan Muslim League became prime minister. In 1993 elections, Bhutto defeated Sharif and became prime minister once again. In 1996, however, following months of allegations of corruption, Pakistan's president again dismissed Bhutto's government.

Recent developments

In 1999, General Pervez Musharraf led a military coup that overthrew Pakistan's democratically elected government. Musharraf dissolved the parliament, suspended the Constitution, and declared himself the head of a transitional government. In a national referendum in April 2002, voters approved the extension of Musharraf's term as president for five years. Musharraf allowed parliamentary elections to be held in October 2002. He was reelected president in 2007.

In December 2007, Benazir Bhutto was assassinated while campaigning for the 2008 parliamentary elections. During the elections, Bhutto's party, the PPP, won the most seats, and Yousaf Raza Gilani was elected prime minister. In August, Musharraf resigned as president. Asif Ali Zardari, Bhutto's widower, was elected president the next month. In 2010, Pakistan's government approved a constitutional amendment that stripped away many of the powers that had been granted to the office of president under Zia and Musharraf.

In 2011, American military forces killed al Qa`ida leader Osama bin Laden in Abbottabad, a city northeast of Islamabad. Bin Laden was believed to have masterminded the September 11 terrorist attacks against the United States.

ENVIRONMENT

From its snowy peaks in the northern frontier to its southern shore on the Arabian Sea, Pakistan is a land of great geographical and climatic contrasts.

The country can be divided into four major regions: (1) the Northern and Western Highlands, (2) the Punjab and Sindh plains, (3) the Balochistan Plateau, and (4) the Thar Desert. Each region contains its own important land features.

The Northern and Western Highlands

Some of the most rugged mountains in the world cover northern and western Pakistan. The Karakoram Range stretches across Pakistan's northeast frontier. The Hindu Kush range extends across Pakistan's northwestern border. Travel through this area is difficult and dangerous.

The Karakoram Range is part of the Himalaya mountain system. It stretches across the northern part of Pakistan. The Karakoram contains the world's second highest peak, K2. It rises 28,250 feet (8,611 meters) above sea level.

The high valleys of the Karakoram Range are home to small, isolated communities. Each of these mountain communities has its own culture and language. About 3,000 Kalash people, for example, live in a border region near the town of Chitral. Unlike other Pakistanis, the Kalash never converted to Islam. They still worship their ancient gods.

Before Pakistan became an independent nation, the people of these mountain communities were much more politically and culturally independent. Now that the Karakoram Highway links Chitral with such communities as Swat and Dir, more trade has come to this area.

Mountain passes cut through the peaks of the Karakoram and the Hindu Kush at several points. One of the most famous is the Khyber Pass, which links Afghanistan and Pakistan.

These mountain regions have the coolest temperatures in Pakistan. Summer temperatures average about 75° F (24° C), and winter temperatures often fall below freezing.

The Punjab and Sindh plains

The Punjab and Sindh plains cover most of eastern Pakistan. The Indus River flows through this region. Its land is called an *alluvial plain,* because it was created from soil deposits left by rivers.

The Indus and its four tributaries—the Chenab, Jhelum, Ravi, and Sutlej—water the Punjab in the north. These rivers meet in east-central Pakistan and flow together as the broadened Indus through the Sindh Plain and southwest to the Arabian Sea.

The plains receive very little rainfall. The eastern part of the Punjab receives the most rain— more than 20 inches (51 centimeters) per year. Rain is brought by the moist summer *monsoon* (seasonal wind), which blows across Pakistan from July to September.

A dense deodar forest blankets the hills behind the palace of the former ruler of Swat in Khyber Pakhtunkhwa Province. The deodar is a durable cedar tree of the Himalaya.

The Bolan Pass links the plains of the Indus River with Quetta, capital of the province of Balochistan. The people of Balochistan raise goats and sheep on the arid land.

Boatmen in Sindh navigate one of the many channels of the Indus River. These channels have been a major transportation system for centuries.

Due to extensive irrigation systems along the Indus and its tributaries, the Punjab and Sindh plains have become major agricultural centers.

The Balochistan Plateau

The Balochistan Plateau lies in southwestern Pakistan. Because it is dry and rocky, plant life is sparse. The people who live in Balochistan tend flocks of sheep and goats on the arid rangeland. The Balochistan Plateau is the driest area in Pakistan. It receives less than 5 inches (13 centimeters) of rain per year.

The Thar Desert

The Thar Desert is located in southeastern Pakistan. Although much of the desert is sandy wasteland, irrigation projects near the Indus River have made the land more suitable for farming.

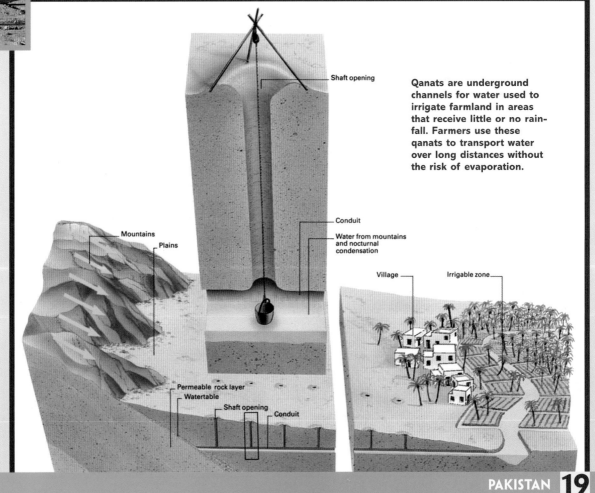

Shaft opening

Qanats are underground channels for water used to irrigate farmland in areas that receive little or no rainfall. Farmers use these qanats to transport water over long distances without the risk of evaporation.

Mountains

Plains

Conduit

Water from mountains and nocturnal condensation

Village

Irrigable zone

Permeable rock layer

Watertable

Shaft opening

Conduit

PEOPLE

When Pakistan became an independent nation in 1947, its population was about 30 million. Today, there are about 173 million people in Pakistan. This dramatic rise in population is due to the nation's high birth rate combined with a lowered death rate that has resulted from improved medical care.

Overpopulation puts a heavy burden on Pakistan. The country does not grow enough food to feed its people, and so it must import much of its food. In addition, Pakistan struggles to meet the demands for housing, health care, education, and electricity.

Language and cultural groups

Language is an important factor that distinguishes Pakistan's cultural groups. Urdu has been Pakistan's national language since independence, but less than 10 percent of the people speak it as their primary language. Major regional languages include Balochi, Punjabi, Pashto (also called Pakhto), and Sindhi. Arabic remains the language of Islam, and English is the everyday language of the upper class.

The Punjabis, Pakistan's largest cultural group, live mainly in the Punjab but have a presence in other parts of the country, especially Karachi. Members of this group control Pakistan's government, economy, and armed forces. Urdu-speaking Muhajirs immigrated to Pakistan from India when the two countries separated in 1947. The Muhajirs became prominent in government after independence, but they have since lost power.

Other leading groups include the Sindhis, the Pashtuns (also called Pakhtuns), and the Balochi. The Sindhis have a slight majority in Sindh but are outnumbered by Muhajirs and other non-Sindhi groups in major cities, such as Karachi. Muhajirs and Sindhis have clashed over political control of Karachi. The Pashtuns, who belong to various tribes and speak Pashto, inhabit the Khyber Pakhtunkhwa Province and the northern part of Balochistan. The Balochi consist of several nomadic and tribal groups. They speak dialects of Balochi and live in Balochistan. Balochistan is also home to smaller groups, such as the Brahuis, Makranis, and Lassis.

A Kalash girl wears the traditional headdress of her people—a cap of colored beads. The Kalash are a small community in the mountains of the Chitral region.

Muslim women wear long robes called *burqas*. The robe covers the woman from the top of her head to the ground except for a veiled opening for the eyes.

People from Afghanistan form another cultural group in Pakistan. Millions of them fled to Pakistan to escape conflicts in Afghanistan.

Religion

Islam shapes the social and religious life of Pakistanis. Even with their many different cultural groups, Pakistanis have a common ground in the religion of Islam. About 97 percent of Pakistanis are Muslims. Most belong to the Sunni branch of the religion, but there is a small minority of Shiah Muslims as well.

For most Pakistanis, religious duties are an important part of everyday life. These include praying five times a day, fasting during the holy month of Ramadan, and making a pilgrimage to Mecca. Midday prayers on Friday mark the Islamic holy day.

City and country life

About two-thirds of Pakistan's people live in rural villages, farming the land or herding for a living. Islamic traditions guide the lives of the rural people. For example, women have far less social freedom than men. Most villagers live in clusters of two- or three-room houses made of clay or sun-dried mud bricks. Many of these houses lack basic plumbing and electric power.

In the cities, most people are unskilled laborers, factory workers, shopkeepers, or craftworkers. They live in small houses in old, crowded neighborhoods. Most city workers follow the same Islamic customs as villagers. Many middle- and upper-class Pakistanis, some of whom have been educated in Europe and the United States, have adopted Western styles and ideas.

Muslims rejoice at a religious festival in Peshawar. Most Pakistanis are Sunni Muslims and follow Islamic tradition in their everyday life.

A polo match is played on a dusty outdoor field. The game originated in Persia about 4,000 years ago. The modern version developed in the Punjab in 1862.

ECONOMY

Although industrial production has increased significantly since the country was created in 1947, Pakistan is still primarily an agricultural country. About half of Pakistan's workers are farmers who still use simple tools to till the land.

Foreign aid and investment have been very important in developing Pakistan's economy. The government has used these funds for a variety of developmental programs. Pakistani leaders have worked to modernize the country's farming methods and equipment. They have also expanded power facilities in the Indus Basin.

Agriculture

The Indus Valley in the Punjab province is Pakistan's main agricultural center. Wheat, cotton, rice, sugar cane, chickpeas, and fruits and vegetables are grown in the region's fertile soil. Most of these crops are used to feed the population.

An extensive irrigation network—the largest in the world—makes farming possible in the Punjab. This network of canals links the Jhelum, Chenab, and Ravi rivers and regulates their flow. Due to this system, otherwise barren land yields two crops per year.

The government encourages farmers to use fertilizers, pesticides, and new types of seeds. These new methods have increased their harvest. Even with the government's attempts to modernize farming methods, crop production is still limited by old-fashioned ways. Many farmers still use basic tools and teams of oxen or buffalo to work the land.

Goats and sheep are raised in regions of Pakistan that are unsuitable for farming. Most of the wool from sheep is exported. Poultry farms are also common in most parts of the country.

Industry

Industrial growth has been rather slow in Pakistan. When the country first became independent, it had few factories. Growth has been limited by lack of money, natural resources, and technology. Today, manufacturing industries employ about 15 percent

Crops of barley, grapes, plums, and wheat are grown in the fertile valley of Hunza in the northern tip of Pakistan. Hunzukuts, many of whom live more than 90 years, believe their long lives are a result of the mineral-rich mountain water.

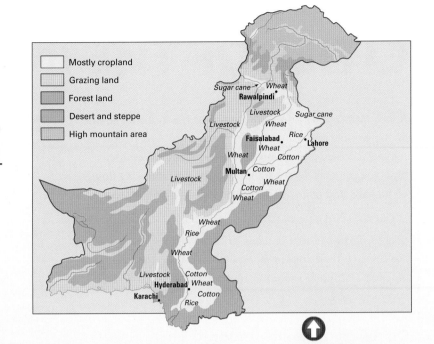

Mostly cropland

Grazing land

Forest land

Desert and steppe

High mountain area

Sugar cane   Wheat
Rawalpindi
Livestock   Sugar cane
Livestock   Wheat
Faisalabad   Rice
Wheat   Lahore
Wheat   Cotton
Multan   Cotton
Livestock   Wheat
Cotton
Wheat

Wheat
Rice

Wheat

Livestock   Cotton
Hyderabad   Wheat
Karachi   Cotton
Rice

The fertile soil of Pakistan's broad alluvial plain yields wheat, cotton, rice, sugar cane, and other crops.

of the population. Leading manufactured products include clothing, fertilizer and other chemical products, food products, and textiles.

The textile industry processes cotton grown on Pakistani farms. Yarn and fabrics are now among the country's leading exports. In addition to processing cotton, Pakistani mills also manufacture artificial silk and synthetic fibers. The food products industry is based on the processing of flour, sugar, and other foods.

Pakistan is one of the world's leading exporters of carpets. The country has a rich tradition of carpet weaving, and Pakistani carpets are prized by collectors all over the world.

Buffalo and oxen, still used as work animals on many Pakistani farms, go to market in Quetta, Balochistan. They also provide meat, milk, and hides.

Pakistani craft workers are famous for their fine carpets. Pakistan is one of the world's leading exporters of carpets.

Newly picked cotton is ready for transport to a processing plant. Cotton grown on the plains is the basis for Pakistan's textile industry.

CITIES OF THE INDUS

About 4,600 years ago, in the valley of the Indus River, one of the world's earliest civilizations was born. The people of the Indus Valley civilization built brick buildings, streets, drainage systems, and warehouses. They created clay figures of humans and animals. They even had a writing system.

Less than 800 years later, this great civilization disappeared. Today, all that remains of these historic people are the ruins of their ancient cities, Mohenjo-Daro (also spelled Moen jo Daro) and Harappa. These ruins tell a remarkable story.

Discovery of a civilization

The ruins of the Indus Valley civilization lay undiscovered until the 1920's, when archaeologists found the remains buried in large mounds. The ruins showed that the Indus civilization covered an area about 1,000 miles (1,600 kilometers) long, extending from the Himalaya to the Indian Ocean.

At first, archaeologists dug mainly at the sites of Mohenjo-Daro on the Indus River and Harappa on the Ravi River, a tributary of the Indus. Later, several hundred smaller settlements were discovered.

An advanced culture

Despite a distance of 342 miles (550 kilometers) apart, Mohenjo-Daro and Harappa were quite similar. They show that a sophisticated culture planned each city very carefully.

The Indus people laid out their streets in a rectangular pattern. Some of the brick houses that lined these streets had elaborate courtyards. These dwellings had a private water supply. Drainage and sanitation systems have also been found.

The Indus people developed a form of writing that was engraved on small stone tablets used as seals. Unfortunately, scholars have so far been unable to translate this writing. As a result, the remains of the Indus Valley civilization leave many questions unanswered.

Archaeologists have concluded that the people of the Indus Valley civilization were farmers and herders. The ruins show that the Indus people stored their grain in large warehouses. They probably traded goods with the people of Mesopotamia, central Asia, southern India, and Persia.

Well-preserved stone steps lead to the remains of brick dwellings at Mohenjo-Daro. Archaeologists uncovered the ruins of this ancient Indus city during the 1920's.

The remains of Mohenjo-Daro include the "Great Bath" (1), which may have been a place of worship as well as an area for washing. A Buddhist stupa (dome-shaped monument) (2), built centuries later, crowns the citadel (3), which also contains a grain warehouse (4). At the southern edge of the city stood a great pillared hall (5).

This statue of a woman was found at Mohenjo-Daro. Indus craft workers made many figures of animals and humans from clay, silver, and bronze.

The figure of a ram is seen in the simple pottery toy that was no doubt enjoyed by a child of the Indus Valley civilization.

Indus art

Archaeologists have unearthed pots, pans, and other utensils made of copper, bronze, and silver in the Indus Valley. The Indus people apparently used some gold, but only for jewelry and decorations.

Many pieces of carved bone and ivory, like those used to decorate furniture, have also been found. From these pieces, scholars believe that making decorated furniture may have been an important craft.

Indus sculptors made clay, bronze, and stone figures of animals and human beings. These may have been used in magic ceremonies. Many of the characteristics of Indus sculpture appear later in Indian sculpture. These include an emphasis on harmonized forms and the contrast of linear rhythm with square and triangular shapes to produce movement.

A people lost to the ages

Scholars are unsure how the Indus Valley civilization began and how it ended. It is astonishing that such an advanced culture could arise from prehistoric farming communities. Why Mohenjo-Daro and Harappa were gradually abandoned is also a mystery.

There is evidence that a natural disaster, such as prolonged flooding, severely damaged the Indus region. Scientists have also noted changes in the courses of the rivers. If such a change left the Indus people without a reliable water supply, they may have become too weak to fight invaders.

Although the reason may never be known, the Indus Valley civilization disappeared by about 1700 B.C.

PALAU

Palau lies about 500 miles (800 kilometers) east of the island of Mindanao in the Philippines. Palau, also spelled *Belau,* is part of the Caroline group of islands in the region of the Pacific referred to as Micronesia. Palau has a population of only about 20,000 people. Peleliu, one of the southern islands in the group, has only about 600 people.

Land

The Palau group consists of a chain of islands surrounded by a coral reef. The islands extend about 100 miles (160 kilometers) from north to south and about 20 miles (32 kilometers) from east to west, and they have a land area of 177 square miles (459 square kilometers).

The northern islands are *volcanic islands,* made of lava built up from the ocean floor by eruptions of underwater volcanoes. Many trees, as well as tropical fruits and vegetables, grow on these fertile islands. The low, flat southern islands of the group are coral islands, consisting chiefly of upraised coral reefs, the limestone formations composed largely of the remains of tiny sea animals. Some of these islands are too rugged for people to live on.

History and people

Archaeological evidence suggests that the Palau Islands were one of the first island groups in Micronesia to be settled. Ancestors of the islanders probably arrived from Asia thousands of years ago. About three-fifths of the people of Palau live in Koror, the largest city, and most work for government agencies. Most of the rest of Palau's people are farmers who live in rural villages. They grow only enough food for their families.

Palau belonged to Germany before World War I (1914–1918) and was turned over to Japan after the war. The Japanese used the islands as their headquarters for all Micronesia. They built roads and concrete piers, developed harbors, and brought in Japanese settlers. On Peleliu, the Japanese dug caves in the soft coral rock for use in defense. In 1935, Palau and other Japanese possessions among the Pacific islands were closed to foreigners.

FACTS

Official name:	Beluu er a Belau (Republic of Palau)
Capital:	Melekeok
Terrain:	Varying geologically from the high, mountainous main island of Babelthuap to low, coral islands usually fringed by large barrier reefs
Area:	177 mi² (459 km²)
Climate:	Wet season May to November; hot and humid
Main rivers:	N/A
Highest elevation:	Mount Ngerchelchuus, 794 ft (242 m)
Lowest elevation:	Pacific Ocean, sea level
Form of government:	Republic in free association with the United States
Head of state:	President
Head of government:	President
Administrative areas:	16 states
Legislature:	Olbiil Era Kelulau (Parliament) consisting of the Senate with 9 members serving four-year terms and the House of Delegates with 16 members serving four-year terms
Court system:	Supreme Court, Land Court, Court of Common Pleas
Armed forces:	The United States is responsible for Palau's defense
National holiday:	Constitution Day - July 9 (1979)
Estimated 2010 population:	20,000
Population density:	113 persons per mi² (44 per km²)
Population distribution:	80% urban, 20% rural
Life expectancy in years:	Male, 69; female, 74
Doctors per 1,000 people:	1.6
Birth rate per 1,000:	12
Death rate per 1,000:	8
Infant mortality:	16 deaths per 1,000 live births
Age structure:	0-14: 24%; 15-64: 70%; 65 and over: 6%
Internet users per 100 people:	27
Internet code:	.pw
Languages spoken:	English and Palauan (official in nearly all states); Sonsoralese and English (official in Sonsoral); Tobi and English (official in Tobi); Angaur, Japanese, English (official in Angaur); Filipino, Chinese, Carolinian
Religions:	Roman Catholic 41.6%, Protestant 23.3%, Modekngei 8.8%, other Christian 6.8%, other 19.5%
Currency:	United States dollar
Gross domestic product (GDP) in 2008:	$173 million U.S.
Real annual growth rate (2008):	N/A
GDP per capita (2008):	$8,238 U.S.
Goods exported:	Copra, shellfish, tuna
Goods imported:	Food, machinery, transportation equipment
Trading partners:	Japan, Philippines, Singapore, United States

Dense woods and sparkling beaches make Palau an attractive vacation spot.

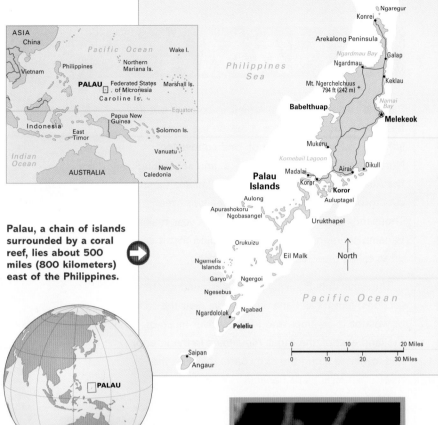

ASIA
China
Pacific Ocean
Wake I.
Vietnam
Philippines
Northern Mariana Is.
PALAU
Federated States of Micronesia
Marshall Is.
Caroline Is.
Equator
Indonesia
East Timor
Papua New Guinea
Solomon Is.
Indian Ocean
AUSTRALIA
Vanuatu
New Caledonia

Palau, a chain of islands surrounded by a coral reef, lies about 500 miles (800 kilometers) east of the Philippines.

PALAU

Ngaregur
Konrei
Arekalong Peninsula
Ngardmau Bay
Galap
Ngardmau
Philippine Sea
Keklau
Mt. Ngerchelchuus 794 ft (242 m)
Namai Bay
Babelthuap
Melekeok
Mukeru
Komebail Lagoon
Madalai
Airai
Oikull
Palau Islands
Koror
Koror
North
Aulong
Auluptagel
Apurashokoru
Ngobasangel
Urukthapel
Orukuizu
Ngemelis Islands
Eil Malk
Garyo
Ngergoi
Pacific Ocean
Ngesebus
Ngardololok
Ngabad
Peleliu
0   10   20 Miles
0   10   20   30 Miles
Saipan
Angaur

Coconuts yield a dried meat called copra, one of the chief exports of Palau. In his right hand, a Palau islander holds a husked coconut seed—a ball of crisp, white, sweet-tasting meat covered by a tough, brown skin and a shell.

In 1944, during World War II (1939–1945), U.S. forces drove the Japanese from the southern Palau islands. After the war ended, Japanese settlers in Palau were sent back to Japan. In 1947, the United Nations established the Trust Territory of the Pacific Islands, under the administration of the United States. The trust territory consisted of about 2,100 islands and atolls in the western Pacific, including Palau.

In 1978, the United States agreed to grant the Caroline Islands a form of self-government called *free association*. The agreement divided the islands into two groups—the Palau Islands and the Federated States of Micronesia. In 1986, the Federated States became a self-governing political unit in free association with the United States. On Oct. 1, 1994, the Republic of Palau followed suit. Later that year, it became a member of the United Nations. In 2006, Palau's capital was moved from Koror to Melekeok.

Under free association, residents of Palau may live and work in the United States. Palau has received hundreds of millions of dollars in assistance from the United States. In return, the United States can use the islands for military purposes. The United States also protects Palau against attack.

A woodcarver puts the finishing touches on a sign for a Palauan community organization. Such traditional crafts are an important part of Palauan culture.

PANAMA

Panama is a small Central American country, but it has worldwide importance. It lies on the narrow Isthmus of Panama, a strip of land connecting North America and South America, and separating the Atlantic and Pacific oceans. The Panama Canal cuts through the isthmus, linking the two oceans. Thousands of ships use the canal each year, making Panama so important to world shipping that it is often called the *Crossroads of the World.*

The canal cuts Panama into eastern and western sections. Almost all Panamanians live near the canal or west of it. Swamps and jungles cover much of the land east of the canal. Mountains in the interior and lowlands along the two coasts make up the three main regions of Panama.

The Central Highland is the mountainous interior. There, the Tabasará Mountains extend eastward from the Costa Rican border, decreasing in height until they are just low hills near the Panama Canal. The mountain valleys provide good farmland. East of the canal, the San Blas Mountains and the Darién Mountains rise to about 6,000 feet (1,800 meters).

The narrow coastal lowlands lie along the Atlantic and Pacific coasts. The Atlantic coast is often called the Caribbean coast because it borders the part of the Atlantic called the Caribbean Sea. The western Pacific Lowland has much fertile farmland.

According to its Constitution, Panama is a republic. The people elect a president to a five-year term, and the Cabinet helps the president run the country. The National Assembly makes the country's laws. Its members are also elected to five-year terms.

The government of Panama has not always followed the Constitution, however. At times, army officers have taken control of the government. In 1989, the United States invaded the country to arrest General Manuel Noriega, a military dictator. Noriega had become increasingly involved in drug trafficking and had declared Panama's 1989 presidential elections invalid when his opponent won. Noriega surrendered to U.S. authorities in January 1990. He completed his U.S. prison sentence in 2007 and was

FACTS

Official name:	Republica de Panama (Republic of Panama)
Capital:	Panama City
Terrain:	Interior mostly steep, rugged mountains and dissected, upland plains; coastal areas largely plains and rolling hills
Area:	29,157 mi² (75,517 km²)
Climate:	Tropical maritime; hot, humid, cloudy; prolonged, rainy season (May to January), short dry season (January to May)
Main rivers:	Tuira
Highest elevation:	Volcán Barú, 11,401 ft (3,475 m)
Lowest elevation:	Pacific Ocean, sea level
Form of government:	Republic
Head of state:	President
Head of government:	President
Administrative areas:	9 provincias (provinces), 3 comarcas (territories)
Legislature:	Asamblea Nacional (National Assembly) with 71 members serving five-year terms
Court system:	Corte Suprema de Justicia (Supreme Court of Justice), superior courts, courts of appeal
Armed forces:	None
National holiday:	Independence Day - November 3 (1903)
Estimated 2010 population:	3,511,000
Population density:	120 persons per mi² (46 per km²)
Population distribution:	72% urban, 28% rural
Life expectancy in years:	Male, 74; female, 79
Doctors per 1,000 people:	1.5
Birth rate per 1,000:	20
Death rate per 1,000:	5
Infant mortality:	15 deaths per 1,000 live births
Age structure:	0-14: 30%; 15-64: 64%; 65 and over: 6%
Internet users per 100 people:	23
Internet code:	.pa
Languages spoken:	Spanish (official), English
Religions:	Roman Catholic 84%, Protestant 15%, other 1%
Currency:	Balboa
Gross domestic product (GDP) in 2008:	$23.20 billion U.S.
Real annual growth rate (2008):	8.3%
GDP per capita (2008):	$6,835 U.S.
Goods exported:	Bananas, fish product, meat, melons, pineapples
Goods imported:	Chemicals, food, machinery, motor vehicles, petroleum products
Trading partners:	China, Colombia, Costa Rica, United States

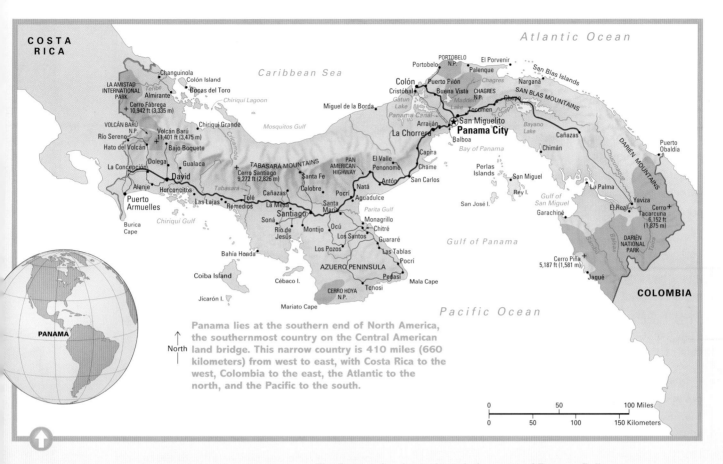

COSTA RICA

Caribbean Sea

Atlantic Ocean

Changuinola
Colón Island
LA AMISTAD INTERNATIONAL PARK
Almirante
Bocas del Toro
Cerro Fábrega + 10,942 ft (3,335 m)
Teribe
Chiriqui Lagoon
VOLCÁN BARÚ N.P.
Chiriqui Grande
Mosquitos Gulf
Volcán Barú + 11,401 ft (3,475 m)
Río Sereno
Bajo Boquete
Hato del Volcán
Dolega
Gualaca
La Concepción
David
TABASARÁ MOUNTAINS
PAN AMERICAN HIGHWAY
El Valle
Penonomé
Alanje
Horconcitos
Cerro Santiago 9,272 ft (2,826 m)
Santa Fe
Antón
Chame
Capira
Puerto Armuelles
Las Lajas
Remedios
Tolé
Cañazas
Colobre
Pocrí
Natá
Aguadulce
San Carlos
La Mesa
Santa Maria
Burica Cape
Chiriqui Gulf
Soná
Santiago
Parita Gulf
Monagrillo
Chitré
Tabasara
Río de Jesús
Montijo
Ocú
Los Santos
Guararé
Bahía Honda
Los Pozos
Las Tablas
Pocrí
Coiba Island
AZUERO PENINSULA
Pedasí
Mala Cape
Cébaco I.
CERRO HOYA N.P.
Tonosí
Jicarón I.
Mariato Cape

PORTOBELO N.P.
Portobelo
El Porvenir
Palenque
Colón
Puerto Pilón
Narganá
San Blas Islands
Cristóbal
Buena Vista
CHAGRES N.P.
Chagres
SAN BLAS MOUNTAINS
Miguel de la Borda
Gatún Lake
Panama Canal
Madden Lake
Gorgona
Chepo
La Chorrera
Arraiján
San Miguelito
Panama City
Balboa
Bay of Panama
Bayano Lake
Cañazas
Chimán
DARIÉN MOUNTAINS
Puerto Obaldía
Perlas Islands
San Miguel
Rey I.
La Palma
Yaviza
Chucunaque
El Real
Cerro Tacarcuna + 6,152 ft (1,875 m)
San José I.
Gulf of San Miguel
Garachiné
DARIÉN NATIONAL PARK
Cerro Piña 5,187 ft (1,581 m) +
Balsas
Sambú
Tuira
Jaqué

Gulf of Panama

COLOMBIA

Pacific Ocean

North

PANAMA

Panama lies at the southern end of North America, the southernmost country on the Central American land bridge. This narrow country is 410 miles (660 kilometers) from west to east, with Costa Rica to the west, Colombia to the east, the Atlantic to the north, and the Pacific to the south.

| 0 | 50 | 100 Miles |
| 0 | 50 | 100 | 150 Kilometers |

sent to France. In France, he was sentenced for money-laundering. France extradited Noriega to Panama in 2011, where he began serving time for crimes committed while he ruled Panama.

The United States has played a role in Panama's history in other ways. Panama was a province of Colombia until 1903, when the United States encouraged the people to revolt and form their own country. The United States then built the Panama Canal and established the Panama Canal Zone. Many Panamanians opposed the United States; some demonstrated and rioted. In 1968, General Omar Torrijos Herrera became a dictator in Panama. Torrijos signed a treaty with the United States that resulted in the transfer of the Canal Zone to Panama in 1979 and the transfer of the canal to Panama on Dec. 31, 1999.

In 1999, Mireya Moscoso became the first woman to be elected president of Panama. She was replaced in 2004 by Martín Torrijos Espino, son of the dictator Herrera. In presidential elections in 2009, voters chose conservative businessman Ricardo Martinelli.

The Panama Canal cuts through the center of Panama. Built by the United States, it links the Atlantic and Pacific oceans. The canal is bordered on both sides by the Panama Canal Zone, a strip of land given to the United States in 1903 but returned to Panama in 1979. In 1999, the United States handed over control of the Panama Canal to Panama.

PEOPLE AND ECONOMY

Panama has a racially mixed population of some 3-1/2 million people. The first inhabitants of what is now Panama were American Indians. In the 1500's, Spanish explorers became the first Europeans to arrive. Spanish settlers soon began to bring Africans to Panama to work as slaves. Later, many more people of African descent came to Panama from Caribbean islands.

Through the years, these groups intermarried. Today, about 70 percent of Panamanians are descendants of more than one group. People of African descent alone make up about 15 percent, people of only European descent make up about 10 percent, and people of *indigenous* (American Indian) descent are about 5 percent of the population.

Spanish is Panama's official language, but many people speak English, too. Some indigenous people speak their own native language in addition to Spanish.

Daily life

Most of Panama's wealthy citizens live near the Panama Canal, a bustling center of urban activity. A small group of wealthy Panamanians, most of them of European descent, are called the *elite*. The elite control the country's economic and political systems. The group includes landowners, doctors, lawyers, and political and military leaders whose families have had wealth for several generations. They take pride in their traditions and tend to avoid contact with less privileged Panamanians.

Many other Panamanians of European and mixed descent are merchants, government officials, and office workers in the Canal Zone and belong to the middle class. Most Panamanians of African descent also live near the canal, but they are mainly poor laborers. As in many other countries, the Africans in Panama suffer from job discrimination.

Away from the canal, Panama is mainly a land of quiet rural areas— farms, tiny villages, and small towns.

Most Panamanians who live away from the canal are farmers. Many must struggle to produce even enough food for their own use.

The main indigenous groups are the Chocó, the Cuna, and the Guaymí. They live mostly in rural areas, where they farm and fish for a living.

The economy

The Panama Canal is the most important single factor in the country's economy. Near the canal, workers are involved in business generated by the waterway—commerce, trade, and manufacturing.

The Panama Canal Authority, a Panamanian government agency, collects tolls from ships that use the canal. The canal also provides jobs for many Panamanians. Some workers operate or maintain the canal, and others work in stores or other businesses connected with the activity around the canal.

Fort San Lorenzo, near Portobelo, is now a playground for some Panamanian children. These young dancers are descended from Africans who came to Panama from Caribbean islands.

Rural Panama is the site of cattle ranches and large plantations, as well as small farms. Many farmers are squatters, who settle on land owned by others and farm it.

A Cuna woman wears a bright and beautifully embroidered headscarf and a ring through her nose. The Cuna, or San Blas Indians, who live on the San Blas Islands off Panama's northern coast, are noted for their distinctive, colorful clothing and jewelry.

Passengers board a public bus in Panama City, the country's capital and largest city. Such buses are an essential form of transportation in the city.

Bananas contribute to Panama's export earnings, along with melons and pineapples. The country's fishing industry produces shrimp and anchovetta.

Commerce and trade flourish near the canal. Panama City and Colón are banking centers, and hundreds of import and export companies operate in Colón.

About two-thirds of Panama's manufacturing companies are located just west of the canal. Chief products include beer, cement, cigarettes, processed foods, and petroleum products.

Away from the canal, the economy is based on agriculture. More Panamanians work in agriculture than in any other activity—about a fourth of the nation's workers.

Most farmers work small plots of land using old-fashioned methods and produce only *subsistence crops*—food for their own families. Rice is the main subsistence crop, followed by corn and beans.

Bananas, sugar cane, coffee, and tobacco are *cash crops*—crops raised for sale. They are grown mainly on large plantations owned by wealthy landowners.

Panama also has small fishing and mining industries. Shrimp and *anchovetta*, a small fish that is ground into fish meal, are the main catches. Mines produce building materials such as lime, sand, and crushed stone. Panama also has large copper deposits, but they are undeveloped.

THE PANAMA CANAL

The Panama Canal is one of the world's greatest engineering achievements. The canal is an artificial waterway that cuts about 50 miles (82 kilometers) across Central America to link the Atlantic and Pacific oceans. It enables thousands of ships each year to travel from one ocean to the other without sailing around South America, thus saving a voyage of more than 7,800 miles (12,600 kilometers).

Three sets of water-filled chambers called *locks* raise and lower ships from one level to another. The locks, which look like giant steps, were built in pairs so that ships can pass in both directions at the same time.

More than half the cargo that passes through the canal is on its way to or from ports in the United States. Another major user of the canal is China.

People dreamed of a canal through Central America for hundreds of years. As early as 1517, the famed Spanish explorer Vasco Núñez de Balboa, who was then a colonial governor, saw the possibility of a canal connecting the Atlantic and Pacific oceans.

In the 1800's, Panama became a province of Colombia. During the 1849 California gold rush, many prospectors sailed from the East Coast of the United States to the Isthmus of Panama, crossed it on mule and on foot, then sailed to California. In 1850, Colombia allowed New York business executives to build the Panama Railroad across the isthmus.

In the late 1800's, a French company owned the rights to build a canal across Panama but failed in its attempt to build a sea-level canal. The rights and property were later offered for sale to the U.S. government.

In 1902, the U.S. Congress gave President Theodore Roosevelt permission to accept the French offer if Colombia would give the United States permanent use of a canal zone. Colombia agreed, but it held out for more money. The Panamanians, with the encouragement of France and the United States, revolted.

On November 3, 1903, Panama declared its independence from Colombia. About two weeks later, Panama and the United States signed a treaty giving the United States permanent, exclusive use and control of a canal zone 10 miles (16 kilometers) wide.

The greatest obstacle to building the canal was disease. The Isthmus of Panama was then one of the most disease-ridden places in the world. Colonel William C. Gorgas, an American physician, took charge of improving health conditions and launched a campaign to destroy the mosquitoes that carried malaria and yellow fever. The first two years of canal building were devoted largely to clearing the brush and draining the swamps where mosquitoes swarmed. Rats, which carried bubonic plague, were eliminated.

The actual construction of the canal included three major engineering feats: (1) the Gaillard Cut had to be made through hills, requiring the removal of millions of cubic yards of soil; (2) a dam had to be built across the Chagres River; and (3) locks had to be built to move ships between different water levels. Engineers believed a canal with locks would be cheaper and faster to build than a sea-level canal. At the height of the construction in 1913, more than 43,000 people worked on the canal. On Aug. 15, 1914, the S.S. *Ancon* became the first ship to travel through the new Panama Canal.

The United States transferred ownership of the canal to Panama in 1999. In 2006, Panamanians voted to expand the Panama Canal so that it could handle more traffic. Construction began in 2007.

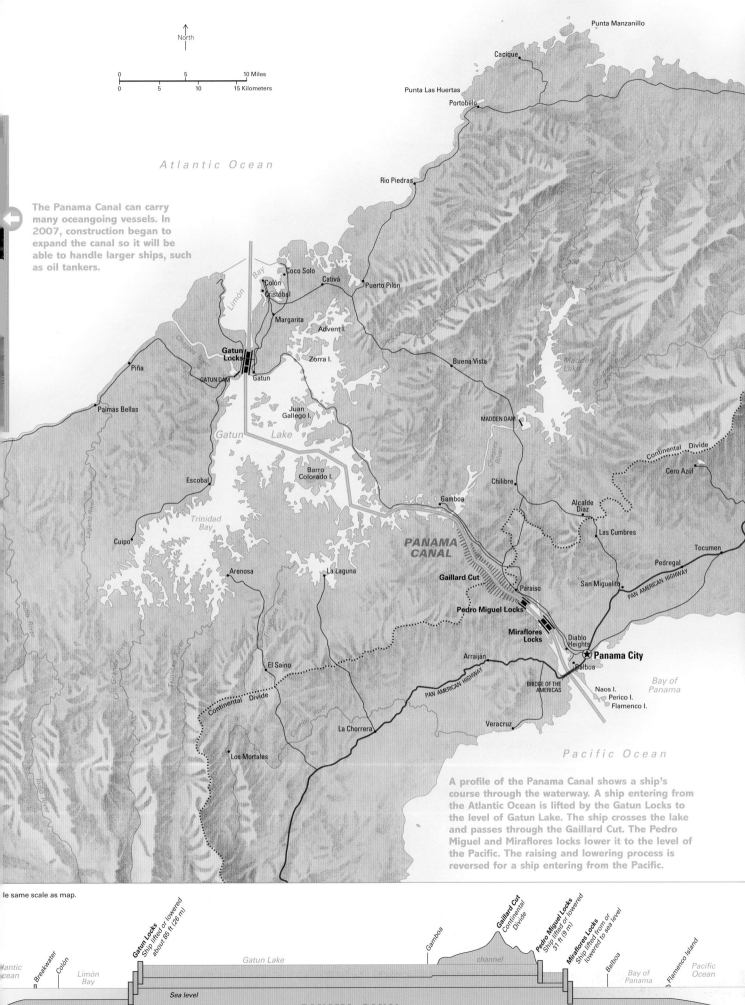

North

0 5 10 Miles
0 5 10 15 Kilometers

Atlantic Ocean

The Panama Canal can carry many oceangoing vessels. In 2007, construction began to expand the canal so it will be able to handle larger ships, such as oil tankers.

Punta Manzanillo
Cacique
Punta Las Huertas
Portobelo
Rio Piedras

Coco Solo
Colón
Cativá
Cristóbal
Puerto Pilón
Margarita
Advent I.
Gatun Locks
Gatun
GATUN DAM
Piña
Zorra I.
Buena Vista
Limón Bay
Chagres River
Palmas Bellas
Juan Gallego I.
Madden Lake
MADDEN DAM
Gatun Lake
Cero Azúl
Continental Divide
Escobal
Barro Colorado I.
Chilibre
Chagres River
Alcalde Díaz
Gamboa
Las Cumbres
Cuipo
Trinidad Bay
La Laguna
PANAMA CANAL
Tocumen
Pedregal
Gaillard Cut
Paraíso
San Miguelito
PAN AMERICAN HIGHWAY
Arenosa
Ciri Grande River
Trinidad River
Pedro Miguel Locks
Indio River
El Saíno
Miraflores Locks
Diablo Heights
Camito River
Arraiján
Balboa
★ **Panama City**
Continental Divide
PAN AMERICAN HIGHWAY
Veracruz
BRIDGE OF THE AMERICAS
Naos I.
Bay of Panama
La Chorrera
Perico I.
Flamenco I.
Los Mortales

Pacific Ocean

A profile of the Panama Canal shows a ship's course through the waterway. A ship entering from the Atlantic Ocean is lifted by the Gatun Locks to the level of Gatun Lake. The ship crosses the lake and passes through the Gaillard Cut. The Pedro Miguel and Miraflores locks lower it to the level of the Pacific. The raising and lowering process is reversed for a ship entering from the Pacific.

le same scale as map.

Gatun Locks
Ship lifted or lowered about 85 ft (26 m)

Gaillard Cut
Continental Divide

Pedro Miguel Locks
Ship lifted or lowered 31 ft (9 m)

Miraflores Locks
Ship lifted from or lowered to sea level

Breakwater
Colón
Limón Bay
Gatun Lake
Gamboa
channel
Balboa
Bay of Panama
Flamenco Island
Pacific Ocean
Atlantic Ocean
Sea level

PANAMA CANAL

Papua New Guinea is an independent nation in the Pacific Ocean just north of Australia. It consists of the eastern part of the island of New Guinea plus a chain of tropical islands that extends more than 1,000 miles (1,600 kilometers). Papua New Guinea shares New Guinea, the world's second largest island, with Papua, a province of Indonesia.

Papua New Guinea has a population of nearly 7 million. About 98 percent of the people are Melanesians. Other groups on the islands include people of Chinese, European, and Polynesian origin. Most of the people live in small rural villages. They farm the land and grow most of their own food.

About 40 percent of the people live in valleys in the interior. People of the interior maintain many of their traditions, including the popular *sing-sings*—celebrations featuring exotic masks and costumes, body painting, music, and dance.

The people of Papua New Guinea speak about 850 languages but manage to communicate with one another by using several widely understood languages. About half of the country's adults can read and write. More than 80 percent of primary-school age children attend school, but only about 15 percent of secondary-school age children do so.

People lived in what is now Papua New Guinea at least 50,000 years ago. The earliest settlers probably migrated from the Asian mainland by way of the Malay Peninsula and Indonesia. Farming in New Guinea began about 9,000 years ago. In the early 1500's, Spanish and Portuguese explorers landed on the islands. The Dutch and English visited several of the islands over the next 300 years.

In 1884, Germany took control of the northeastern part of New Guinea, and the United Kingdom took the southeastern part. The United Kingdom turned over administration of its territory to Australia in 1905.

Australia seized the German region during World War I (1914–1918), and the League of Nations placed all of New Guinea under Australian rule after the war. During World War II (1939–1945), Japanese forces in-

FACTS

Official name:	Independent State of Papua New Guinea
Capital:	Port Moresby
Terrain:	Mostly mountains with coastal lowlands and rolling foothills
Area:	178,704 mi² (462,840 km²)
Climate:	Tropical; northwest monsoon (December to March), southeast monsoon (May to October); slight seasonal temperature variation
Main rivers:	Fly, Purari, Sepik
Highest elevation:	Mount Wilhelm, 14,793 ft (4,509 m)
Lowest elevation:	Pacific Ocean, sea level
Form of government:	Constitutional monarchy
Head of state:	British monarch, represented by governor general
Head of government:	Prime minister
Administrative areas:	20 provinces
Legislature:	National Parliament or House of Assembly with 109 members serving five-year terms
Court system:	Supreme Court
Armed forces:	3,100 troops
National holiday:	Independence Day - September 16 (1975)
Estimated 2010 population:	6,719,000
Population density:	38 persons per mi² (15 per km²)
Population distribution:	87% rural, 13% urban
Life expectancy in years:	Male, 59; female, 64
Doctors per 1,000 people:	Less than 0.05
Birth rate per 1,000:	30
Death rate per 1,000:	10
Infant mortality:	50 deaths per 1,000 live births
Age structure:	0-14: 40%; 15-64: 58%; 65 and over: 2%
Internet users per 100 people:	2
Internet code:	.pg
Languages spoken:	Pidgin English, English, Police Motu (all official); about 850 indigenous languages
Religions:	Protestant 69.4%, Roman Catholic 27%, other (including Bahai and indigenous beliefs) 3.6%
Currency:	Kina
Gross domestic product (GDP) in 2008:	$8.09 billion U.S.
Real annual growth rate (2008):	6.3%
GDP per capita (2008):	$1,294 U.S.
Goods exported:	Coffee, copper, crude oil, gold, palm oil, wood products
Goods imported:	Food, machinery, petroleum and petroleum products, transportation equipment
Trading partners:	Australia, China, Japan, Singapore

vaded New Guinea in 1942. Heavy fighting then took place on New Guinea and some of the other islands.

Australia set up a new government after the war. It granted Papua New Guinea control over its internal affairs in 1973, and in 1975 Papua New Guinea gained complete independence.

Papua New Guinea is now a parliamentary democracy. It is also a member of the Commonwealth of Nations, an association of independent countries and other political units under British law and government.

As one of the independent members of the Commonwealth of Nations, Papua New Guinea recognizes the British monarch as head of the Commonwealth, and a governor general represents the British monarch on the islands. But the monarch is mainly a symbol, with no real power to govern. The people elect a national legislature, which selects a prime minister to head the government.

Port Moresby is the capital and largest city of Papua New Guinea. The city, which lies on a deep harbor on the southeastern coast of the island, is served by an international airport and has a university and other educational institutions.

Between 1988 and 1998, conflict over mining and land rights on the island of Bougainville turned into a rebellion aimed at independence. A cease-fire was signed in 1998. Papua New Guinea granted greater *autonomy* (self-rule) to Bougainville in a peace agreement in 2001. Bougainville elected its first autonomous government in 2005.

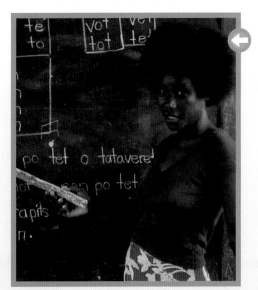

Pidgin English (Tok Pisin) and Police Motu, as well as English, are widely understood languages in Papua New Guinea. In addition, about 850 local languages are spoken.

Port Moresby lies on the hot, humid coast of southeastern New Guinea. The nation's capital has many modern buildings, but most people live in houses built on stilts.

Papua New Guinea lies in the Pacific Ocean. It consists of the eastern part of the island of New Guinea plus a chain of tropical islands that extend more than 1,000 miles (1,600 kilometers).

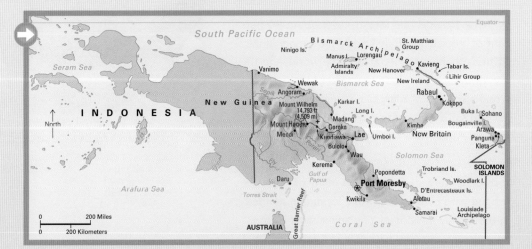

LAND AND ECONOMY

Papua New Guinea lies only a few degrees south of the equator, and just north of Australia. It has a total land area of 178,704 square miles (462,840 square kilometers). The eastern half of the island of New Guinea makes up most of the country. The rest of the country consists of the islands of the Bismarck Archipelago, Bougainville and Buka in the Solomon Islands chain, the D'Entrecasteaux Islands, the Louisiade Archipelago, the Trobriand Islands, and Woodlark Island.

The country has a hot, humid climate. Temperatures average from 75° F to 82° F (24° C to 28° C) in the lowlands and about 68° F (20° C) in the highlands, with an average annual rainfall of about 80 inches (203 centimeters).

New Guinea's animals include crocodiles, tree kangaroos, and such snakes as the death adder, the Papuan black, and the taipan. The island also has many brilliantly colored birds and butterflies.

Volcanoes, rain forests, and coral

The country's larger islands, including New Guinea, New Britain, and Bougainville, have many high mountain ranges. Volcanoes are common on their northern coasts, and the islands experience frequent, sometimes severe, earthquakes. These mountainous islands are located in a zone, called the *Ring of Fire,* that encircles the Pacific Ocean. The majority of the world's volcanoes are found along the Ring of Fire.

Dense tropical rain forests cover about 80 percent of the islands. Rain forests have more kinds of trees than any other area in the world, and they are always green. In addition, more than half the world's species of plants and animals live in tropical rain forests. Swamps cover much of the coastal land of Papua New Guinea.

The country's outlying small islands are the tops of underwater mountains. Many of them are fringed with coral, a limestone formation molded in the sea by millions of tiny animals. When the animals die, they leave limestone "skeletons" that

A government worker shows local farmers how to plant tree saplings that will renew the forests. When mature, the new trees will replace those that have been cut down to clear land for cultivation.

High mountain ranges cross New Guinea from east to west. Few people live on these rugged, forested slopes. Most of the island's people live in farming communities.

A New Guinean hunter carries his equipment with him to a festival.

New Guinea's coastal areas consist mainly of swamps and clumps of mangroves. Off the coast lie coral reefs filled with colorful sea animals.

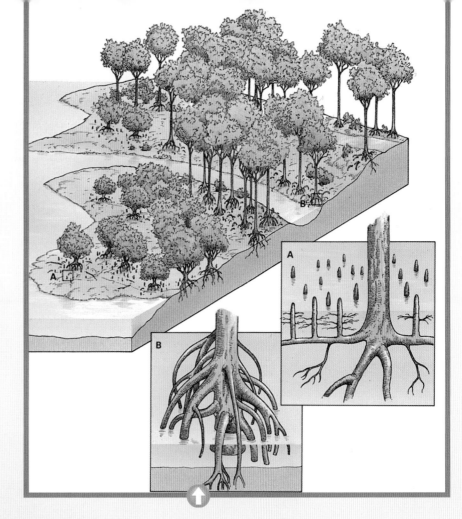

A member of another ethnic group has his face painted with vegetable dyes in preparation for a ceremony.

Mangroves grow along coasts in salty tropical waters. Mangroves colonize coastal areas by sending out pioneer roots (A) that trap mud. Thousands of stiltlike roots (B) anchor the trees. They also help build shorelines by catching and holding particles of dirt and sand.

form the foundations of barriers and ridges in the sea called *coral reefs*. Some coral reefs protect harbors, but many endanger shipping.

The economy

Most of the people in Papua New Guinea raise crops for a living. They grow most of their own food, including sweet potatoes, yams, taro, and the potatolike root called *cassava*. They also produce products for sale, including cocoa, coconuts, coffee, palm oil, rubber, and tea.

Copper, gold, and petroleum are the most valuable exports. Gold and copper are mined near Papua New Guinea's border with Indonesia. Gold mines also operate at Porgera

in the New Guinea central highlands and on the island of Lihir. Bougainville, the largest of the Solomon Islands, has rich copper and gold deposits, but political unrest in the late 1900's interrupted mining operations there.

The islands have sawmills that are used to process timber. Logs, sawed timber, plywood, veneers, and wood chips are exported to Australia, New Zealand, Asia, and Europe.

The rugged terrain of Papua New Guinea has been an obstacle in its economic development. The dense forests, steep mountains, and swampy coasts make transportation difficult and expensive. Building and maintaining roads is difficult. Many parts of the country can only be reached by air, foot, or canoe. A national airline operates among the islands.

PARAGUAY

Paraguay is a small, landlocked country near the center of South America. It is surrounded by Argentina, Bolivia, and Brazil.

As a country rich in natural resources—including fertile soil, dense forests, and vast hydroelectric power potential—Paraguay has many opportunities for economic growth. However, as a result of political instability and wars with neighboring nations, Paraguay is a poor country with a developing economy. Most Paraguayans belong to the lower class. Nearly half the people are farmers, and most grow barely enough food to feed their families.

About 95 percent of Paraguay's people are *mestizos*—people of mixed European and *indigenous* (native American Indian) ancestry. They are descendants of the Guaraní Indians, who intermarried with Spanish settlers. The Guaraní lived in what is now Paraguay long before the first Spaniards arrived in the 1500's.

Early Spanish settlements

Spanish and Portuguese explorers first came to Paraguay in search of a shipping route westward across the continent to the gold and silver mines of Peru. In 1537, Juan de Ayolas traveled up the Paraná and Paraguay rivers to a point north of what is now the capital city of Asunción. Part of his expedition stayed behind to establish a settlement at Asunción. The settlement became the seat of government for all of Spain's colonies in the southeastern part of South America.

The Guaraní offered little resistance to the Spaniards, and the territory became a Spanish colony. In 1588, Jesuit priests began to arrive in Paraguay to convert the Guaraní to Roman Catholicism. The Jesuits organized mission settlements called *reducciones,* or *reductions,* where many Guaraní lived and worked. The Guaraní received food, clothing, and other goods in exchange for tending cattle and working in the fields.

By the 1730's, there were about 30 reducciones in Paraguay, with a total population of about 140,000. The reducciones—which had become quite prosperous as a result of their exports of cotton, tobacco, yerba maté, hides, and wood—also protected the Guaraní from Portuguese slave traders and Spanish colonists who wanted to use them for cheap labor.

The success of the reducciones aroused the envy of the Spanish settlers, who complained to the Spanish king, Charles III. In 1767, King Charles expelled the Jesuits from Paraguay. Groups of settlers looted the reducciones, and the native people either

A rural Paraguayan proudly wears a handwoven sash in his country's national colors during Independence Days, which are celebrated on May 14 and 15. Like most Paraguayans, he is a mestizo (a person of mixed white and indigenous ancestry).

A shopkeeper pauses for a refreshing cup of yerba maté, a favorite beverage of Paraguayans. Urban dwellers like this shopkeeper generally have a higher standard of living than the rural people because of educational and job opportunities. Medical services are also more readily available in the cities. Many city people work in government, business, or the professions, or as craft workers, unskilled laborers, or factory workers. Just over half the people in Paraguay live in cities and towns.

A woman buys melons at a summer market in Asunción, the country's capital and largest city. Melons are a popular summer fruit in Paraguay, along with oranges, pineapples, and watermelons.

returned to their former way of life or went to work on the large colonial estates. In 1776, Paraguay became part of the Viceroyalty of La Plata, a large colony created by Spain out of its territories in southeastern South America.

Independence and after

In 1811, Paraguay declared its independence from Spain. Since then, the country's history has been dominated by dictatorships or near-dictatorships. The first dictator was José Gaspar Rodriguez de Francia (known as *El Supremo*). He was chosen to serve as president by the new national assembly in 1814 and declared dictator for life two years later. Because Francia distrusted foreigners, he kept Paraguay isolated from the rest of the world.

After Francia's death in 1840, Carlos Antonio López ruled Paraguay as dictator. López reversed Francia's isolationist policies and encouraged trade with foreign nations. His son, Francisco Solano López, succeeded him in 1862. Solano López provoked serious quarrels with Argentina, Brazil, and Uruguay. Hostilities eventually led to the War of the Triple Alliance in 1865—a conflict that left Paraguay in ruins when the war finally ended five years later.

In the 1930's, a war with Bolivia over the Chaco led to some territorial gain, but Paraguay also suffered many casualties.

PARAGUAY TODAY

In May 1989, Paraguay held its first free elections in 35 years. The people elected General Andrés Rodríguez Pedotti as president. Earlier that year, Rodríguez had led the overthrow of President Alfredo Stroessner, Paraguay's dictator since 1954. After Stroessner's fall from power, Paraguayans danced in the streets of Asunción, celebrating the end of what was perhaps the most brutal and corrupt dictatorship in Paraguay's history.

The Stroessner era

In 1954, General Alfredo Stroessner, who was commander in chief of Paraguay's armed forces, led a military overthrow of President Federico Chaves's government. Between 1954 and 1988, Stroessner was reelected seven times, but the elections were controlled by the police and military forces.

Paraguay's Constitution allowed the president widespread powers, including the power to dissolve the national legislature. Stroessner used these powers, along with military and police power, to gain absolute control of the government.

Stroessner allowed little opposition to his rule, imprisoning some Paraguayans who criticized his policies and sending others into exile. Stroessner's secret police maintained an army of spies among the population. The spies reported anybody who complained about the government, and ordinary people lived in constant fear of being turned in to the authorities by a jealous neighbor.

Stroessner brought political stability to Paraguay, but at the expense of civil and human rights. Because he was able to maintain a stable government, he attracted foreign aid and investment to Paraguay. Stroessner used the money to begin a broad program of economic development that included modernizing agriculture, building roads, and promoting industry. But in the end, his programs benefited only a small number of Paraguayans—mostly Stroessner, his associates, and the wealthy landowners.

FACTS

Official name:	Republica del Paraguay (Republic of Paraguay)
Capital:	Asunción
Terrain:	Grassy plains and wooded hills east of Paraguay River; Gran Chaco region west of Paraguay River mostly low, marshy plain near the river, and dry forest and thorny scrub elsewhere
Area:	157,048 mi² (406,752 km²)
Climate:	Subtropical to temperate; substantial rainfall in the eastern portions, becoming semi-arid in the far west
Main rivers:	Paraguay, Paraná, Tebicuary, Verde
Highest elevation:	2,231 ft (680 m) near Villarrica
Lowest elevation:	180 ft (55 m), at the meeting point of the Paraguay and Paraná rivers
Form of government:	Constitutional republic
Head of state:	President
Head of government:	President
Administrative areas:	17 departamentos (departments), 1 capital city
Legislature:	Congreso (Congress) consisting of the Camara de Senadores (Chamber of Senators) with 45 members serving five-year terms and the Camara de Diputados (Chamber of Deputies) with 80 members serving five-year terms
Court system:	Corte Suprema de Justicia (Supreme Court of Justice)
Armed forces:	10,700 troops
National holiday:	Independence Day - May 14 (1811) Note: observed May 15
Estimated 2010 population:	6,502,000
Population density:	41 persons per mi² (16 per km²)
Population distribution:	59% urban, 41% rural
Life expectancy in years:	Male, 71; female, 76
Doctors per 1,000 people:	1.1
Birth rate per 1,000:	27
Death rate per 1,000:	6
Infant mortality:	25 deaths per 1,000 live births
Age structure:	0-14: 36%; 15-64: 59%; 65 and over: 5%
Internet users per 100 people:	9
Internet code:	.py
Languages spoken:	Spanish (official), Guarani (official)
Religions:	Roman Catholic 89.6%, Protestant 6.2%, other Christian 1.1%, other 3.1%
Currency:	Guarani
Gross domestic product (GDP) in 2008:	$16.11 billion U.S.
Real annual growth rate (2008):	5.5%
GDP per capita (2008):	$2,537 U.S.
Goods exported:	Cotton, leather, meat products, soybeans, tobacco, vegetable oils, wood products
Goods imported:	Chemicals, electronics, machinery, motor vehicles, petroleum products
Trading partners:	Argentina, Brazil, China, United States, Uruguay

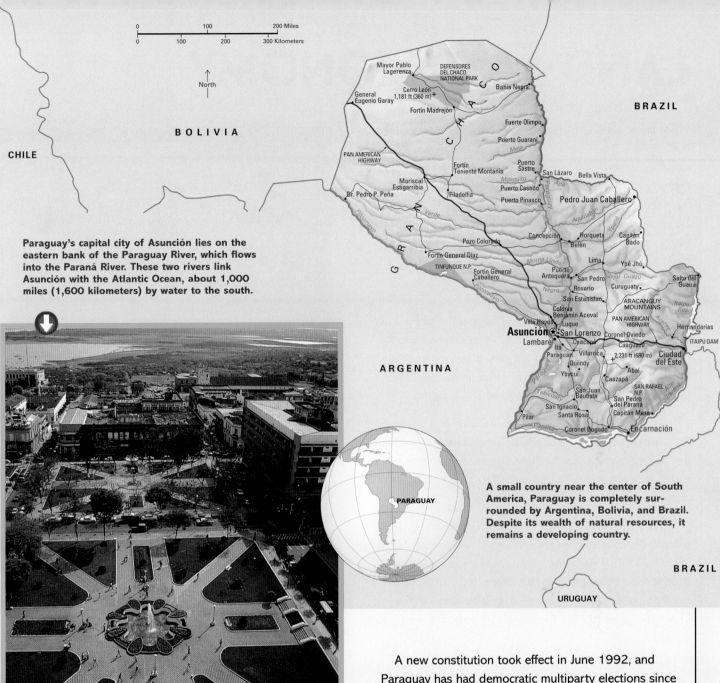

Paraguay's capital city of Asunción lies on the eastern bank of the Paraguay River, which flows into the Paraná River. These two rivers link Asunción with the Atlantic Ocean, about 1,000 miles (1,600 kilometers) by water to the south.

A small country near the center of South America, Paraguay is completely surrounded by Argentina, Bolivia, and Brazil. Despite its wealth of natural resources, it remains a developing country.

Return to democracy

General Rodríguez's rise to power closely paralleled Stroessner's. Like Stroessner, Rodríguez was a military officer who seized control by overthrowing the government and holding elections before opposing political parties had a chance to form. Rodríguez was also a member of the Colorado Party, which had controlled the voting booths at all of Stroessner's reelections.

Even so, the 1989 elections were the first in 35 years that were not predetermined in favor of the Colorado Party. Rodríguez was elected president.

A new constitution took effect in June 1992, and Paraguay has had democratic multiparty elections since 1993. The Colorado Party's candidate won the presidential elections in 1993, 1998, and 2003.

In 1999, Paraguay's vice president was assassinated. The National Congress blamed President Raúl Cubas Grau and former army chief Lino Oviedo Silva for planning the assassination. Cubas resigned, and both men fled the country. Senate President Luis González Macchi then succeeded to the presidency of Paraguay and served until the next election.

In 2008, the election of Fernando Lugo as president marked the end of about 60 years of government by the Colorado Party. Lugo was backed by the center-left Patriotic Alliance for Change.

LAND AND ECONOMY

Paraguay is divided by the Paraguay River into two major land regions. The Chaco, officially called Occidental Paraguay, stretches westward from the Paraguay River. It is a region of coarse grasses, scrub forests, thorny shrubs, and salt marshes. Eastern Paraguay, officially called Oriental Paraguay, lies east of the Paraguay River. A fertile region of rolling hills and thick forests, Eastern Paraguay is home to more than 95 percent of the country's population.

The Chaco

A large region that occupies about three-fifths of Paraguay, the Chaco is part of the Gran Chaco that extends into Argentina and Bolivia. Covered with dry grass and sparsely dotted with quebracho trees and other hardwoods, the Chaco is a desolate, undeveloped region.

Paraguay fought for control of the Chaco in a war with Bolivia that began in 1932. Paraguay suffered many casualties during the war, but the fighting ended in a truce in 1935. A final settlement that gave Paraguay nearly all the disputed territory was reached in 1938.

Several slow-moving rivers flow through southern and eastern Chaco, including the Pilcomayo and the Verde. The Pilcomayo forms Paraguay's southwestern border with Argentina. Like other rivers in the Chaco, the Pilcomayo often overflows after heavy summer rains. Some rivers in the Chaco disappear during the dry winter season, forming salt marshes.

The underground water throughout much of the Chaco is too salty for drinking or irrigation, making the land unsuitable for farming. The quebracho trees that grow in the region are harvested for *tannin,* which is used in tanning hides.

Most of the Chaco is uninhabited. The few people who live in the Chaco include cattle ranchers in the south, a small group of German-speaking people of the Mennonite faith in the central part, and scattered tribes of Guaraní Indians in the most remote sections.

Eastern Paraguay

In contrast to the Chaco, Eastern Paraguay is the productive heartland of the country. Its rich soil yields such crops as *cassava* (a root vegetable), corn, cotton, rice, soybeans, sugar cane, tobacco, and wheat. The thickly forested Paraná

The setting sun shimmers in the waters of Lake Ypacaraí, which lies near the Paraguay River. Waterways are important transportation routes in Paraguay, and hydroelectric power plants on the Acaray and Paraná rivers provide plentiful electricity.

Workers clear the way for a steam locomotive to change tracks. Paraguay has only one major railway system. It runs between the capital city of Asunción and the city of Encarnación in the southeastern corner of the country.

The poorest city dwellers live in shantytowns in houses made of wood or metal scraps. But Paraguay does not have the large urban slums typical of some other Latin American countries.

Plateau occupies about a third of the region. Asunción, the nation's capital and largest city, lies in Eastern Paraguay.

Most of the people in Eastern Paraguay live along the Paraguay River or in the southwestern part of the region, where small towns and farming villages dot the countryside. Most rural Paraguayans live in *ranchos,* one-room houses with a dirt floor and walls made of reed, wood, or brick. A separate or attached shed serves as a kitchen. Few ranchos have indoor plumbing.

Economic development

Service industries and agriculture, including forestry, are the most important sectors of the Paraguayan economy. About half the country's workers are employed in service industries, especially at banks, government agencies, health care facilities, schools, stores, restaurants, and hotels.

Agriculture, which consists mainly of cattle ranching in the Chaco and crop growing in Eastern Paraguay, employs about a third of the workers. Since the mid-1900's, the government has encouraged farmers to buy land in undeveloped areas. However, only about a third of the land that could be used for crop production is presently cultivated.

Thick forests, which are valuable for the timber they produce, cover about half of Paraguay. *Deforestation* (the destruction of forests) is becoming a problem, especially in Eastern Paraguay.

Manufactured goods include cement, processed foods, steel, textiles, and wood products. Hydroelectric power plants on the Acaray and Paraná rivers provide plentiful electricity, but Paraguay has no petroleum reserves and must import oil.

A mestizo woman sells oranges at a wayside stall. Paraguayan mestizos, descended from Guaraní Indians who intermarried with Spaniards, still maintain many traditional Guaraní ways. Most Paraguayans speak both Guaraní and Spanish, but the Guaraní language is used in everyday conversation. The popular music of Paraguay also reflects the slow rhythm of traditional Guaraní music.

Surrounded by the rugged, sharply rising peaks of the Peruvian Andes, the famous "lost" city of Machu Picchu clings to a ridge 8,000 feet (2,400 meters) above sea level. Once a royal retreat for the ruling Inca family, Machu Picchu remained hidden from the outside world until it was discovered by Hiram Bingham in 1911. Today, as the cold mountain wind whistles through its stone ruins, Machu Picchu is a silent but awe-inspiring reminder of a once-mighty civilization that was cut down by Spanish invaders from half a world away.

While the Inca created the most famous of Peru's early civilizations, theirs was by no means the first. Scholars believe that the first people to live in Peru came there from North America about 12,000 years ago. The story of Peru, South America's third largest country, began in the small farming villages of these primitive tribes.

Peru's breathtaking scenery is a fitting backdrop for its long, eventful history. Its mountains are taller than Europe's Alps, its deserts are drier than Africa's Sahara, and its people and their story are no less dramatic.

Early civilizations

Peru's earliest people learned to farm, domesticated the llama, and were the first to cultivate the potato, which grew wild in the highlands. The Chavin people developed Peru's first known civilization, which flourished from about 800 to 400 B.C. The Chavin spoke Akaro, which developed into the Aymara language, still spoken by a number of native goups in Peru.

Later, other peoples, such as the Moche (or Mochica), Tiwanaku, and Chimu, developed civilizations in Peru. The Tiwanaku settled near the cold, high shores of Lake Titicaca. Little is known of them except for the massive stone carvings, such as the *Puerta del Sol* (Doorway of the Sun), which they left behind. The remains of the Moche, who settled in northern Peru, and the Nazca of the south contain secrets of their own.

The Moche civilization flourished between the A.D. 100's and 700's. The Moche inhabited the almost rainless coastal strip of Peru but cultivated large areas of farmland by developing irrigation techniques. They were also skillful builders, constructing huge stepped pyramids more than 100 feet (30 meters) high. The Nazca, who lived about the same time as the Moche, etched huge geometrical and animal figures, visible only from the air, into the parched desert landscape.

PERU

The Chimu and Inca

The largest and most important pre-Incan civilization was that of the Chimu, who settled on the coastal plain near the present-day city of Trujillo. The Chimu settlements were among the largest urban areas of the time. They began to build their capital city of Chan Chan about A.D. 1000. Today, the city's ruins cover about 8 square miles (20 square kilometers). Scholars believe the Chimu actually developed much of the political organization, irrigation techniques, and road-building skills for which the Inca later became famous.

The Inca conquered the Chimu in 1471. By the early 1500's, the Inca had built a great empire, and their civilization had reached its peak. The Inca rule extended north into parts of present-day Colombia and Ecuador and south as far as present-day Chile and Argentina.

The Spanish and independence

Spanish adventurer Francisco Pizarro was on a quest for fabled Incan treasures of gold and silver when he entered Peru about 1527. The riches he found then convinced him to return with about 180 men in 1532. With the aid of additional Spanish troops and some native allies, he conquered most of Peru by the end of 1533.

Soon after the conquest, the king of Spain appointed a *viceroy* (governor) to enforce Spanish laws and customs. Enslaving the *indigenous* (native American Indian) people to work in the mines and on colonists' plantations, the Spanish made Peru one of their most profitable colonies. From time to time, indigenous groups and *mestizos* (people of mixed native and Spanish ancestry) rebelled unsuccessfully.

Finally, in the early 1800's, the heroes of Peru's wars of independence—José de San Martín, Simón Bolívar, and Antonio José de Sucre—freed Peru from Spanish rule. Today, though Peru suffers from economic problems and civil strife, the ruins of its ancient cultures stand as an eternal testament to its noble heritage.

45

PERU TODAY

Since independence, Peru has struggled with a variety of political, economic, and social problems.

Social structure

Many of Peru's social challenges stem from the strict class system established so long ago by its Spanish conquerors. Under the original two-class system, a small upper class of Spanish ancestry controlled a huge native population. Some Spaniards and indigenous people married, and their descendants became known as *mestizos*. As the number of mestizos grew, they too became part of the lower class.

During the 1900's, class and racial barriers became less rigid, and a middle class developed. By the early 2000's, the middle class included about half the people of European ancestry and many mestizos. Today, many middle-class mestizos attend college and become leaders in government, industry, the armed forces, and the professions. Some have entered the upper class. But the majority of mestizos and many of Peru's indigenous people—a group that makes up nearly half of the population—remain poor and uneducated. The upper class continues to control most of the nation's wealth.

Struggle for equality

Until the 1920's, most of Peru's political parties favored the upper class. Then, in 1924, Víctor Raúl Haya de la Torre founded the *Alianza Popular Revolucionaria Americana* (American Popular Revolutionary Alliance, or APRA), a party that demanded equal rights for all Peruvians and public ownership of Peru's basic industries.

When Haya de la Torre lost the 1931 presidential election, APRA charged dishonesty in vote counting and staged violent antigovernment demonstrations. Continued violence led the government to outlaw APRA. Although the group's legality was restored in 1956, APRA lost its popularity to Fernando Belaúnde Terry's Popular Action Party. Belaúnde was elected president in 1963 and worked for native rights.

A military group seized control of Peru's government in 1968. It began land redistribution reform and took over some foreign-owned industries. An elected civilian

FACTS

Official name:	Republica del Peru (Republic of Peru)
Capital:	Lima
Terrain:	Western coastal plain, high and rugged Andes in center, eastern lowland jungle of Amazon Basin
Area:	496,225 mi² (1,285,216 km²)
Climate:	Varies from tropical in east to dry desert in west; temperate to frigid in Andes
Main rivers:	Amazon, Marañón, Ucayali
Highest elevation:	Huascarán, 22,205 ft (6,768 m)
Lowest elevation:	Pacific Ocean, sea level
Form of government:	Constitutional republic
Head of state:	President
Head of government:	President
Administrative areas:	25 regiones (regions) and 1 provincia (province)
Legislature:	Congresso Constituyente Democratico (Democratic Constituent Congress) with 120 members serving five-year terms
Court system:	Corte Suprema de Justicia (Supreme Court of Justice)
Armed forces:	114,000 troops
National holiday:	Independence Day - July 28 (1821)
Estimated 2010 population:	28,971,000
Population density:	58 persons per mi² (23 per km²)
Population distribution:	73% urban, 27% rural
Life expectancy in years:	Male, 68; female, 73
Doctors per 1,000 people:	1.2
Birth rate per 1,000:	21
Death rate per 1,000:	6
Infant mortality:	24 deaths per 1,000 live births
Age structure:	0-14: 31%; 15-64: 63%; 65 and over: 6%
Internet users per 100 people:	26
Internet code:	.pe
Languages spoken:	Spanish (official), Quechua (official), Aymara, many Amazonian languages
Religions:	Roman Catholic 81.3%, Evangelical 12.5%, other 6.2%
Currency:	New sol
Gross domestic product (GDP) in 2008:	$127.60 billion U.S.
Real annual growth rate (2008):	9.2%
GDP per capita (2008):	$4,373 U.S.
Goods exported:	Clothing, coffee, copper, fish products, gold, petroleum products, zinc
Goods imported:	Crude oil, food, machinery, motor vehicles, plastics
Trading partners:	Brazil, Canada, Chile, China, Japan, United States

government was reinstated in 1980. During the 1980's two leftist groups—the Shining Path and the Tupac Amaru Revolutionary Movement (MRTA)—began guerilla attacks to overthrow the government. Peru also faced economic problems.

In 1990, independent candidate Alberto Fujimori was elected president. In 1992, he suspended the Constitution and dissolved the legislature, which he said was corrupt. Peruvians approved a new constitution in 1993 that allowed Fujimori to run for reelection in 1995.

Also in 1992, police arrested Abimael Guzmán Reynoso, the leader of Shining Path. He and 10 others were convicted of treason. In late 1996, the MRTA seized the residence of the Japanese ambassador to Peru, taking 490 hostages. The rebels released some of the hostages, and in 1997, Peruvian troops stormed the residence, freeing the rest.

In 2000, after election and corruption scandals, Fujimori went into exile in Japan. He was arrested in 2005 while traveling in Chile and was returned to Peru in 2007. In 2007 and 2009, Fujimori was convicted of corruption and other charges.

In elections in 2001, Alejandro Toledo became Peru's first indigenous president. Alan García Perez of the APRA, who had served as president from 1985 to 1990, was reelected in 2006. In 2011, Peruvians elected left-wing former military officer Ollanta Humala Tasso as president. Humala formed Peru's first democratically elected leftist government.

Peru lies on the west coast of South America, along the Pacific Ocean. With its vast mineral deposits, fertile farmland, and coastal waters teeming with fish, Peru has great economic potential. However, industrial development has been slow, and most Peruvians, especially the indigenous population, live in poverty.

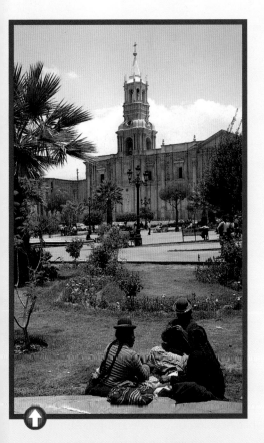

The Plaza de Armas, a spacious central square in the Peruvian town of Arequipa, provides a comfortable spot for women to work on their crafts. The Spaniards built many richly decorated buildings throughout Peru. Some have been destroyed by earthquakes and rebuilt in the traditional Spanish colonial style.

ENVIRONMENT

Highland people watch their herd of llamas outside their thatch-roofed house. These indigenous people live at elevations up to 15,000 feet (4,570 meters) above sea level. The Himalaya in Asia is the only other place on Earth where people live at such high altitudes.

A woman gathers potatoes in a woven basket. A native plant of South America, the potato was cultivated by the Inca about 500 years ago. The Inca freeze-dried potatoes to make a light, floury substance called chuño, from which they made bread.

Peru is a land of great geographical contrasts. From the western coastal deserts to the cool central Andes and the humid eastern forests, the rugged landscape of Peru has had a profound influence on the life of its people. Peruvians in each of these very different regions have adapted their way of life to the unique characteristics of the land and climate around them.

The Pacific coast

Although nearly all of Peru's west coast along the Pacific Ocean is dry desert land, about 50 rivers flow down from the Andes Mountains and cross the region. The river water is used for irrigation, enabling farmers to grow crops in this otherwise barren land. Cotton and sugar cane are grown on these coastal farms.

The waters off the coast provide a rich catch for Peru's fishing industry, including anchovettas, sardines, tuna, and other ocean fish. The anchovettas and sardines are ground into fish meal and exported all over the world for use in livestock feed.

Peru's large cities and factories are also located in the coastal area. Lima, the capital and largest city, is the nation's major commercial, cultural, and industrial cen-

ter. Lima is a bustling, modern city, with high-rise buildings lining the streets of its business district. The city's plants and factories produce beer, clothing, cotton and woolen fabrics, and fish meal.

While Lima's wealthy families live in spacious colonial mansions and luxurious suburban homes, the city's poor live in rundown public housing or crowded slums. About a third of the city's population lives in squatter communities known as *pueblos jóvenes* (young towns).

These new towns sprang up when former slumd-wellers began to settle on public land outside the large cities. There, they built shacks made of cardboard, old

Squatter settlements spread out along the outskirts of Lima. Since the 1950's, large numbers of people have migrated from rural areas to the cities.

metal, and other scrap materials. Because the so-called squatters paid no rent, some were able to save enough money to build a permanent house. To encourage these efforts, the government has supplied some new towns with running water and a sewerage system.

The Andes highlands

East of the coastal regions, the Andes Mountains rise sharply. The indigenous people of the highlands make their homes in the plateaus and valleys between the soaring peaks. They work as farmers and herders, growing coffee on the lower eastern mountain slopes and herding llamas, alpacas, goats, and sheep in the higher valleys and on the Altiplano (plateau) surface.

The people of the highlands live at elevations up to 15,000 feet (4,570 meters). Most of their houses have *adobe* (dried clay) walls and a roof made of grass thatch or handmade tile. Although the young people dress in modern clothing, many of the older native people prefer the traditional garments of their ancestors. The men wear ponchos, leggings tied at the knee, and colorful caps with earflaps. The women wear layers of handwoven, brightly colored skirts and derby hats.

The selva

Indigenous groups also make up the majority of the population in the selva, a region of forests and jungles in eastern Peru. The people belong to about 17 different language families, which are further divided into smaller groups. The houses in their scattered villages are built of sticks or bamboo poles with a thatched roof of grass or palm leaves.

Most of the villagers in the Selva grow a few crops, such as corn and *cassava* (a starchy root), but they hunt and fish for most of their food. The area provides a great variety of fish and small game, as well as several kinds of fruits and nuts.

Carefully tended fields near Cusco extend to the Andean peaks. Once the heartland of the Inca empire, this region is now farmland where local people grow crops such as potatoes and corn. Many farm families produce barely enough food to feed themselves.

RAILROAD ABOVE THE CLOUDS

"Wherever a llama can go, so can my railway!" These were the proud words of Henry Meiggs, the ambitious American engineer who laid the route for Peru's famous Central Railway in the late 1800's. Meiggs was sure that he could build a fast, direct rail link between Peru's huge upland mineral reserves and the port of Callao on the Pacific Ocean, which was a suburb of Lima.

Meiggs died in 1877, long before his project could be completed, and his dream of a railroad that would cross the Andes never quite came true. But although the track ends at Huancayo, Peru's Central Railway remains one of the great engineering marvels of all time.

The world's highest standard-gauge railroad, the Central Railway begins its journey at sea level and climbs to an incredible 15,844 feet (4,829 meters)— higher than Mont Blanc, the tallest mountain in the Alps.

For even the most adventurous traveler, a ride in the railway's yellow-and-orange cars can be a harrowing experience. During the all-day journey, passengers endure extremes of temperature—from Lima's summer heat and humidity to Huancayo's bitter cold. But the reward for the traveler is a rare view of extraordinary scenery in this mountainous land.

The Central Railway's journey begins at Callao, but most passengers join the train in Lima at a station called *Desemparados* (The Forsaken Ones). Passenger services are available on several days each month. During the course of the journey, the train crosses almost 60 bridges and goes through more than 60 tunnels.

The train pulls out of Desemparados past a vast shantytown crowding the banks of the Rímac River. Traveling eastward, the train crosses Peru's narrow, almost rainless coastal plain and passes through a small canyon—barely a cut in the rock—beyond Chosica. Then the train begins its remarkable climb, and high cliffs, often only a few feet from the carriage windows, dominate the view.

A train passes through the Vilcanota Valley on the line that runs between Cusco in the Andean highlands and Puno on Lake Titicaca. Peru's railroads offer unforgettable views of the Andes Mountains.

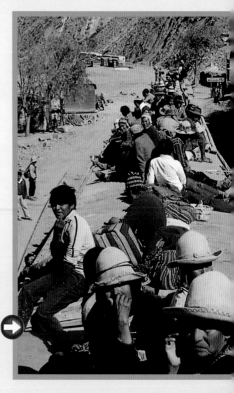

Passengers crowd carriage roofs on a railway line in the Andean highlands.

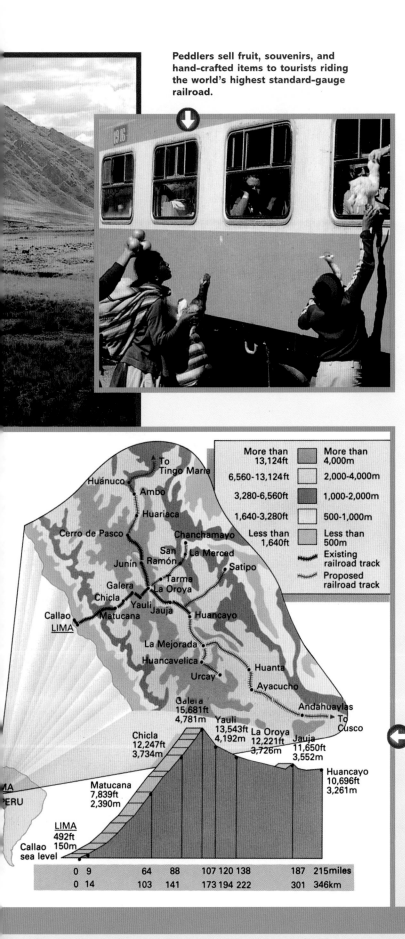

Peddlers sell fruit, souvenirs, and hand-crafted items to tourists riding the world's highest standard-gauge railroad.

On its upward climb, the train seems to defy gravity. In some places, the rise is too steep to allow even zigzag bends. At these points, the train is driven along a ramp beyond the track, and then shifted to the next stretch of track.

By the time the train reaches Matucana, it has climbed to 7,842 feet (2,390 meters) above sea level. At this height, some passengers may begin to feel the effects of *soroche* (altitude sickness). These effects include dizziness, shortness of breath, or nausea. Attendants in white jackets are on hand to operate oxygen tanks for those who need help. Some passengers chew aspirin as they inhale the oxygen, while others rebuild their strength by taking glucose pills. Many native Peruvians chew coca leaves, which may help keep them from feeling hungry or tired but do not nourish the body.

At Ticlio, the line forks. The main branch passes through the Galera Tunnel and then heads on to Huancayo. This branch reaches its highest altitude of 15,688 feet (4,782 meters) inside the Galera Tunnel. Another branch travels from Ticlio to the mining town of Mirococha. The highest point on the Central Railway is reached on a siding at 15,844 feet (4,829 meters) along this branch at La Cima.

The world's highest standard-gauge railroad was built by Henry Meiggs, an American who fled to South America in the 1850's to escape his creditors. In the past, the railroad carried great quantities of gold, silver, lead, and mercury from the mountain mines to the port of Callao. Today, Peruvians and foreign tourists ride the famous train up to Huancayo to enjoy the spectacular views and to buy beautiful hand-crafted items made by indigenous Quechua-speaking people.

THE INCA

According to Inca legends, the Inca civilization began when four brothers and their four sisters emerged from a cave south of what later became Cusco. The brothers and sisters wandered the land for many years, searching for a place to live. Eventually, only Manco Capac and his sister Mama Ocllo were left. They founded the city of Cusco, which later became the center of a great empire.

Originally, the name *Inca* only referred to the members of the empire's royal court. After the Spanish conquest, some people started using the name *Inca* for all the South Americans who were once ruled by the Inca emperors.

Inca descendants celebrate the festival of the sun god with the Dance of the Condors. Most Peruvians belong to the Roman Catholic faith brought by the Spaniards, but many combine Catholic beliefs with indigenous practices.

Early history

The Inca may have settled in the Cusco area around the A.D. 1200's. By about 1400, they ruled one of several powerful kingdoms in the region.

The Inca empire emerged in the early 1400's under the ruler Pachacuti, who is said to have defeated an invading kingdom and built up a large army. Pachacuti, his son Topa Inca Yupanqui, and his grandson Huayna Capac expanded their territory through a combination of alliances and conquests. By the early 1500's, the empire stretched along the west coast of South America for more than 2,500 miles (4,020 kilometers).

An advanced culture

Because the Inca did not develop a system of writing, historians have relied on archaeological remains to learn how they lived. The Inca were accomplished engineers and builders, though they did not use any wheeled carts or vehicles. They constructed a network of elaborate footpath roads and suspension bridges connecting distant regions of their empire. The Inca were also skilled craft workers who created many fine articles from gold and silver.

A Quechua woman displays her earrings, a reminder of the goldsmithing skills of her ancestors. Quechua, still spoken today, was the language of the Inca.

Canals at Tambomachay, Peru still survive as evidence of the remarkable engineering skills of the Inca. In the highland regions, irrigation systems provided water for crops grown on terraced hillsides.

The massive walls of the Inca fortress Sacsahuaman contain some stones that are 16 feet (5 meters) long. Workers hauled the rock from quarries more than 35 miles (56 kilometers) away. Inca builders cut large stone blocks so precisely that they fit together perfectly without cement. Not even a knife blade can fit between them.

Before their defeat at the hands of the Spanish conquistadors in 1532, the Inca ruled a vast empire that covered parts of what are now Colombia, Ecuador, Peru, Bolivia, Chile, and Argentina. The Inca worshiped gold as sacred to the sun god. Inca rulers stored huge quantities of the precious metal and carefully controlled trade in metals and precious stones. Gold objects were used by the families of Inca nobles, and gold figures were placed in many Inca graves.

Colombia
Equator
Ecuador • Quito
Amazon
Cajamarca
Brazil
Peru
Pachacamac
Ollantaytambo
Machu Picchu
Sacsayhuaman
Cusco
Lake Titicaca
Bolivia
INCA EMPIRE
South Pacific Ocean
Chile
MOUNTAINS
Argentina

Inca Empire in 1527
• Inca city

0   400 Miles
0   400 Kilometers

The Inca were a deeply religious people. They believed the world was created by a god called Viracocha. The ruling family prayed chiefly to Inti, the sun god. The Inca also worshiped the earth and the sea as goddesses. They frequently held religious ceremonies in which crops and animals—and sometimes humans—were sacrificed to keep the good will of the gods.

A farming people, the Inca raised crops of corn, cotton, potatoes, an edible root called *oca,* and a grain known as *quinoa.* In the highlands, those who worked the land lived in small houses made of adobe or built of stones set in mud. Coastal villagers built houses out of cane twigs and poles. The nobles lived in spacious, richly decorated stone palaces.

March of the conquistadors

When the Inca emperor Huayna Capac died in about 1527, civil war broke out between rival groups following two of his sons, Atahualpa and Huascar. The fighting between the brothers' followers seriously weakened the empire. In 1532, Atahualpa's army defeated and captured Huascar. Meanwhile, the Spanish conquistador Francisco Pizarro marched into Inca territory with about 180 men.

Pizarro and his men ambushed and captured Atahualpa in the city of Cajamarca, Peru. Although a ransom of a room filled with gold and a room filled twice with silver was paid for Atahualpa's return, Pizarro broke his word and executed his prisoner anyway. While the Spanish held Atahualpa prisoner, Atahualpa's army had executed Huascar. The remaining Inca leaders retreated in shock and confusion and were quickly defeated by the Spaniards.

A weaver displays the colors of the rainbow on an open-air loom near Cusco, the former Inca capital. The native people of the Andean highlands still produce brightly colored woven cloth as their Inca ancestors once did.

LAKE TITICACA

Lake Titicaca lies on the border between Peru and Bolivia. At an altitude of 12,507 feet (3,812 meters) above sea level, it is the world's highest navigable lake. Shrouded in ancient Indian myth and mystery for thousands of years, the lake—resting high on a plateau in the central Andes—is unusual because of its size and high altitude.

Lake Titicaca is the result of water collecting on the plateau instead of flowing down from the mountains into the ocean. The Desaguadero River flows out of the lake's southern end and empties into Lake Poopó in Bolivia. Because the level of Lake Poopó has now fallen below its outlet, water usually can escape only through evaporation. When the water is high, it overflows into nearby swamplands.

In pre-Inca times, Lake Titicaca was the center of an important civilization. Ruins of the ancient ceremonial center of Tiwanaku stand 12 miles (20 kilometers) away. Lake Titicaca, which is about 110 miles (180 kilometers) long and about 45 miles (72 kilometers) wide, covers an area of some 3,200 square miles (8,300 square kilometers). Parts of the lake are more than 900 feet (270 meters) deep.

From steamboats to reed boats

The first steamboat to cross Lake Titicaca was the *Yaravi*, a Scottish-built 200-ton steamer. In 1862, the *Yaravi* sailed from Scotland to the Peruvian coast, where it was completely dismantled. The ship's parts were then hauled up the rugged Andean terrain on the backs of mules. When the mules reached Puno, on the western shore of Lake Titicaca, the *Yaravi* was reassembled.

Today, commercial ferryboats carry passengers and freight across the lake during the day, while smugglers operate in the dark of the night. Slipping past Bolivian Navy patrol boats, these smugglers carry meat, sugar, flour, and coca to Peru. They return to the Bolivian shore with clothes, radios, beer, and textiles. Sharing the waters of Lake Titicaca with the modern motorized boats are *totoras*—small reed boats whose design dates from ancient times.

Lake Titicaca is located on the border between Peru and Bolivia. It is the highest navigable lake in the world.

The totoras captured the interest of the Norwegian ethnologist and author Thor Heyerdahl. Heyerdahl asked the native totora builders to help him build a reed boat. In 1947, Heyerdahl sailed a balsa-wood raft named *Kon-Tiki* from Peru to the Tuamotu Islands in eastern Polynesia to test his theory that the islands of Polynesia could have been settled by native peoples from South America. In 1970, Heyerdahl and a crew of seven sailed a totora reed boat named *Ra II* from Morocco to Barbados in the West Indies. He claimed that this voyage proved that the ancient Egyptians could have sailed similar boats to the New World.

People of the lake

Today, native Aymara and Quechua people inhabit the shores of Lake Titicaca and the surrounding area. But centuries ago, a people called the Uru lived there. Larger Aymara tribes drove the Uru from their lands, and some Uru people fled the mainland to live on artificial islands on the lake. The Uru people have disappeared or merged with other tribes of the region, but some people still make floating-island homes by weaving totora reeds together into huge platforms.

These people eat potatoes, yucca, *oca* (a tuberous root), and a grain called *quinoa*—grown in soil taken from the lakeside and spread over the matted reeds. They also barbecue *cuy*, or guinea pigs.

An Aymara man uses a pole to push his reed boat, known as a totora, through the water. He wears the traditional hat of his tribe—an earflapped, brightly colored woolen hat called a gorro. The slopes of the central Andes rise in the distance.

Modern steamships are now a common sight on the busy waters of Lake Titicaca. For the Inca, the lake was a sacred place.

Totoras, the small reed boats used for centuries in the Lake Titicaca region, are made from totora reeds collected from the shallows of the lake. The green totora leaves are dried in the sun until they are free of moisture and then are bundled up and lashed together with grass to form the boat's hull. The dry reeds float easily on top of the water, but after a few months they become waterlogged and settle lower in the water. To keep their totoras "seaworthy," the owners must periodically dry their boats on shore and launch them again.

Indian children chew on the soft cores of totora reeds as if the reeds were sticks of candy. The totora reed, which the local Indians use to make their floating island homes, is edible and forms an important part of the Indians' diet.

PHILIPPINES

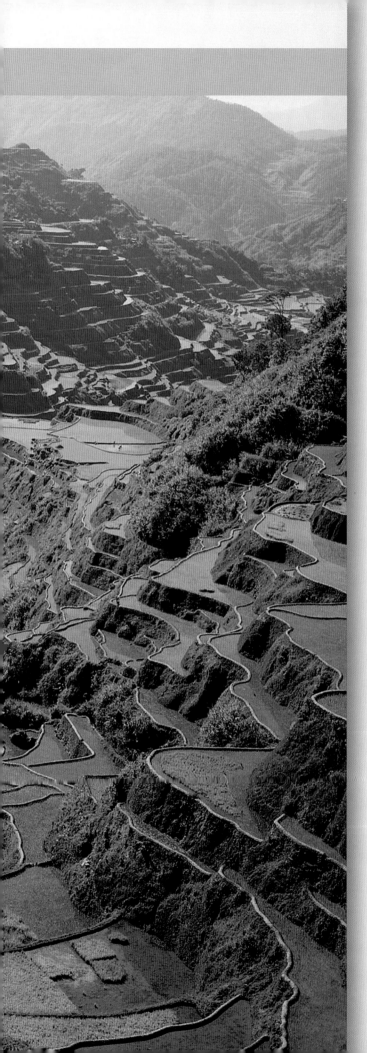

The Philippines is an island country that lies off the Southeast Asian mainland in the Pacific Ocean. People live on about 1,000 of the more than 7,000 islands of the Philippines. Eleven main islands make up more than 95 percent of the country.

Thick tropical forests originally covered most of the Philippines, though lumbering has significantly reduced the forests since the 1950's. Narrow strips of lowland lie along the coasts. Most of the country's larger islands have volcanic mountains, with many active volcanoes. Violent earthquakes occur frequently on the islands. Nature poses an additional threat to the area in the form of *typhoons,* tropical storms that strike seasonally.

The people of the Philippines, called *Filipinos,* have a wide variety of languages, customs, and cultures. Their ancestors migrated from Indonesia and Malaysia thousands of years ago. They formed small communities throughout the area, and each group developed its own culture.

Ferdinand Magellan, a Portuguese sea captain, led a Spanish expedition to the Philippines in 1521. He was looking for a western sea route from Europe to the Spice Islands, now part of Indonesia. Magellan remained in the Philippines for several weeks. He converted many of the people to Christianity before he was killed in a battle between rival Filipino groups.

Another group of Spanish explorers followed in 1565, colonizing the Philippines and naming them after King Philip II of Spain. The Spaniards converted most Filipinos to Christianity, but some tribes kept their own religion. As Christianity spread through the islands, it worked as a unifying cultural force throughout the many scattered communities. Today, more Christians live in the Philippines than in any other Asian nation.

In 1898, as part of the treaty that ended the Spanish-American War, the United States paid Spain $20 million and took control of the Philippines. The United States ruled the islands until the Philippines became a self-governing commonwealth in 1935. During World War II (1939–1945), the Philippines had strategic importance, and it was the scene of extensive, heavy fighting.

The Philippines became an independent nation on July 4, 1946, with a constitution and economic system similar to those of the United States. Since independence, economic inequalities, government corruption, interference with civilian rule by the military, and radical groups on both the right and left have presented the Philippines with its greatest challenges.

PHILIPPINES TODAY

The Philippines is a democratic republic. The people elect the president and vice president to six-year terms. The Congress consists of a 24-member Senate and a House of Representatives with up to 250 members. Up to one-fifth of the representatives are selected from lists drawn up by the political parties to ensure representation of women, ethnic minorities, and certain economic and occupational groups.

The Philippines is divided into 16 regions, each governed by a regional council. The regions, in turn, are divided into 80 provinces. Each province has a governor, a vice governor, and a provincial council elected by the people. The nation has more than 100 *chartered cities,* each headed by an elected mayor. The country also has thousands of *municipalities* (towns) and *barangays* (villages), each governed by elected officials and councils.

Since the late 1990's, the Philippine government has faced a number of challenges to its stability. Several administrations were charged with political corruption.

In addition, Muslim separatist groups have been fighting for independence since the 1970's. In 1996, the government and the largest of these groups signed a peace treaty that created a region of self-rule for Muslims in the southern Philippines. However, during the first decade of the 2000's, fighting continued to erupt between government troops and several Muslim factions.

U.S. base withdrawals

Clark Air Base, a key U.S. military facility about 10 miles (16 kilometers) from Mount Pinatubo, was buried under ash from volcanic eruptions in the early 1990's. As a result, the U.S. government abandoned the base. The United States wanted to continue operations at Subic Bay Naval Station, the largest U.S. naval base abroad, about 25

FACTS

Official name:	Republika ng Pilipinas (Republic of the Philippines)
Capital:	Manila
Terrain:	Mostly mountains with narrow to extensive coastal lowlands
Area:	115,831 mi² (300,000 km²)
Climate:	Tropical marine; northeast monsoon (November to April); southwest monsoon (May to October)
Main rivers:	Agno, Pampanga, Magat, Cagayan, Agusan, Pulangi
Highest elevation:	Mount Apo, 9,692 ft (2,954 m)
Lowest elevation:	Philippine Sea, sea level
Form of government:	Republic
Head of state:	President
Head of government:	President
Administrative areas:	80 provinces, more than 100 chartered cities
Legislature:	Kongreso (Congress) consisting of the Senado (Senate) with 24 members serving six-year terms and the Kapulungan Ng Mga Kinatawan (House of Representatives) with a maximum of 250 members serving three-year terms
Court system:	Supreme Court
Armed forces:	106,000 troops
National holiday:	Independence Day - June 12 (1898)
Estimated 2010 population:	93,715,000
Population density:	809 persons per mi² (312 per km²)
Population distribution:	63% urban, 37% rural
Life expectancy in years:	Male, 67; female, 73
Doctors per 1,000 people:	1.2
Birth rate per 1,000:	26
Death rate per 1,000:	5
Infant mortality:	23 deaths per 1,000 live births
Age structure:	0-14: 35%; 15-64: 61%; 65 and over: 4%
Internet users per 100 people:	6
Internet code:	.ph
Languages spoken:	Filipino (official), English (official), about 70 native languages
Religions:	Roman Catholic 80.9%, other Christian 11.6%, Muslim 5%, other 2.5%
Currency:	Philippine peso
Gross domestic product (GDP) in 2008:	$168.58 billion U.S.
Real annual growth rate (2008):	4.6%
GDP per capita (2008):	$1,880 U.S.
Goods exported:	Clothing, copper, electrical equipment, electronics, food, machinery, transportation equipment
Goods imported:	Chemicals, electronics, machinery, petroleum and petroleum products, transportation equipment
Trading partners:	China, Hong Kong, Japan, Singapore, United States

miles (40 kilometers) south of Pinatubo. In 1991, the United States and the Philippines signed a treaty agreeing to extend the lease on the site for at least 10 years. However, the Philippine Senate voted to reject the treaty because opposition to it was widespread. The United States withdrew from Subic Bay.

Living standards

Economic conditions vary widely among the people of the Philippines. In rural areas, most farmland is divided into large and medium-sized farms. The owners hire laborers who live and work on the land. Most rural houses have wooden walls and roofs made of thatch or corrugated iron. They usually are clustered in small groups. Wealthy families live in large houses surrounded by walls.

In the cities, government-built housing projects are common, but many poor urban people live in roughly-constructed shanties in sprawling slums. Unemployment and underemployment have been serious challenges. Millions of Filipinos work in other countries and send part of their earnings home.

The Rajah Sulayman Park lies in the heart of the Malate district in Manila. The district attracts many visitors for its restaurants, shopping, and entertainment facilities. The Malate Church stands at the rear of the park. A famous "dancing waters" fountain flows in the middle of the park.

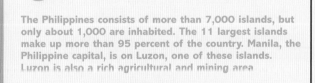

The Philippines consists of more than 7,000 islands, but only about 1,000 are inhabited. The 11 largest islands make up more than 95 percent of the country. Manila, the Philippine capital, is on Luzon, one of these islands. Luzon is also a rich agricultural and mining area.

ENVIRONMENT

The islands of the Philippines extend 1,152 miles (1,854 kilometers) from north to south and 688 miles (1,107 kilometers) from east to west. They lie between several seas that are part of the northwestern Pacific Ocean. The South China Sea and Sulu Sea are to the west, the Celebes Sea to the south, and the Philippine Sea to the east.

The islands contain large regions of dense tropical rain forests. They also have many fine bays and harbors and several large lakes. Most of the rivers flow only in the rainy season, which lasts from June to February. Narrow strips of lowland lie along the coasts.

The Philippine islands lie in a zone of intense earthquake and volcanic activity known as the *Ring of Fire*. This zone runs along the western, northern, and eastern edges of the Pacific Ocean, and more than half of the world's active volcanoes lie within it. In 1991, Mount Pinatubo, a volcano on Luzon, erupted many times, causing more than 800 deaths and destroying crops, homes, and other property.

Off the coast of the island of Mindanao lies the Philippine Trench, one of the deepest spots in all the oceans. It lies 34,578 feet (10,539 meters) below the surface of the Pacific. The ocean's average depth is about 12,900 feet (3,900 meters).

Climate

The Philippines has a hot, humid climate. During the hottest months, from March to May, temperatures may reach 100° F (38° C). The weather cools off during the rainy season, but the temperature rarely falls below 70° F (21° C). Rainfall in the Philippines averages 100 inches (250 centimeters) a year, with some areas receiving up to 180 inches (457 centimeters). Less rain falls on the lowlands than in the high mountain areas because the mountains block winds that carry rain-bearing clouds from the ocean.

About five typhoons strike the Philippines yearly. Between May and November, these storms, with winds of more than 74 miles (119 kilometers) per hour, howl in from the Pacific. Typhoons can rip banana groves to shreds and demolish buildings. Low islands have been completely swamped by the huge waves lashed up by these winds.

Wildlife

A wide variety of plants and animals lives in the Philippines. Banyan and palm trees flourish in the forests, while thick groves of bamboo and thousands of kinds of flowering plants grow throughout the islands. Crocodiles, monkeys, snakes, and many species of tropical birds live in the Philippines. *Tarsiers*—small mammals with large, owl-like eyes—are found only in the tropical rain forests of the Philippines and the East Indies.

Much of the natural environment of the Philippines has been altered for agricultural purposes, such as the cultivation of rice (below). Rice is an essential crop because most Filipinos eat it at every meal.

Resources

The Philippine islands form three groups. The northern group consists of two large islands, Luzon and Mindoro. Luzon produces most of the nation's rice and tobacco, and has large deposits of copper, gold, and other minerals. Manila, the nation's capital, lies on Luzon's southwest coast.

The central group of islands, called the Visayas, consists of about 7,000 islands. Among them are Panay, Cebu, and Bohol, densely populated islands with fertile agricultural land. Cebu, the most crowded of these islands, produces corn, rice, sugar cane, tobacco, and coconuts. Its major city, also called Cebu, is a busy port.

The southern group consists of Mindanao and the Sulu Archipelago, a group of about 400 islands that extend south and west toward Borneo. Mindanao, at the southeastern end of the Philippines, has the country's highest mountains—including some active volcanoes. It is one of the world's leading producers of *abacá,* a plant used in making rope. The island has many fruit plantations.

Typhoons—violent tropical storms that form in the western Pacific—often strike the Philippines. A typhoon is a low-pressure area in the atmosphere around which winds spiral in a counterclockwise direction. Near the center, or "eye," of the storm, wind speeds range from 74 miles (119 kilometers) per hour to 180 miles (290 kilometers) per hour. Lashing rains, violent thunder, and lightning usually accompany the winds.

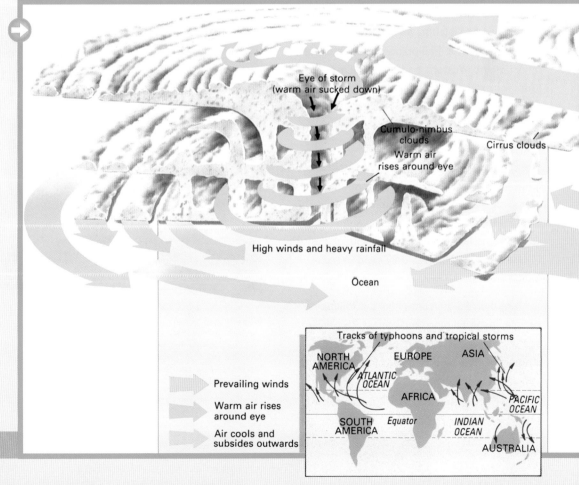

Eye of storm (warm air sucked down)

Cumulo-nimbus clouds

Cirrus clouds

Warm air rises around eye

High winds and heavy rainfall

Ocean

Prevailing winds

Warm air rises around eye

Air cools and subsides outwards

Tracks of typhoons and tropical storms

NORTH AMERICA — EUROPE — ASIA

ATLANTIC OCEAN

AFRICA

PACIFIC OCEAN

SOUTH AMERICA — Equator — INDIAN OCEAN

AUSTRALIA

PEOPLE

The population of the Philippines is about 94 million, and the number has been increasing at a very high rate. The population density in the Philippines is 809 persons per square mile (312 per square kilometer), about three times higher than the average population density in Asia.

A blend of cultures

Anthropologists believe that a tribe of Negritos called the Aeta came to the islands from the Southeast Asian mainland more than 30,000 years ago. They were probably the first people to live in the Philippines. Some descendants of these people still live in isolated mountain areas of the islands.

About 3000 B.C., groups of Malays from Indonesia and Malaysia began to settle along the coasts of the Philippines. Most Filipinos are descended from these Malays.

Chinese make up the second largest group in the Philippines, and smaller numbers of Americans, Europeans, Indians, and Japanese also live on the islands. All these groups have contributed to the Philippine culture, a blend of Asian and Western traditions. Philippine food, for example, is a mixture of American, Chinese, Malay, and Spanish dishes.

Languages

The Philippines has two official languages, Filipino and English. Filipino is the national language, and it is a required subject in all elementary schools. Filipino is a variation of Tagalog, the language of the people of the Manila area. English is widely used in commerce and government.

More than 50 percent of all the Philippine people speak Filipino, and almost 75 percent speak English. In addition, about 70 native languages, all based on Malay languages, are spoken in the Philippines.

A Moro man lives in the Sulu Archipelago in the southern Philippines. The members of this Muslim minority have formed several separatist groups to demand political freedom for their people.

The carabao, the chief domestic animal in the Philippines, gets a holiday in May called Carabao Day. Farmers use the carabao, a type of water buffalo, to pull plows and haul loads.

Filipino girls wear their best clothes to Mass. Most Filipinos follow the Roman Catholic religion, which was introduced by the Spanish in the 1500's.

An elaborate headdress symbolizes this elderly man's high position among his people. He belongs to one of the small tribal groups who live in the remote mountain areas.

Education

About 90 percent of the Philippine people can read and write. The law requires children from 7 to 12 years old to attend school through at least the sixth grade. Teachers in public elementary schools conduct classes in the local dialect for the first two years and then introduce English and Filipino. Most private schools, high schools, and universities use English.

Way of life

Most Filipinos wear Western clothing. On holidays and other special occasions, Philippine men may wear a *barong tagalog,* a beautifully embroidered shirt made of pineapple fiber, raw silk, or cotton. Women may wear a long, puff-sleeved dress called a *balintawak.*

About one-third of the Philippine people live in rural areas, and about two-thirds live in urban areas. More than half the people live on Luzon, the country's largest island. Rural people farm for a living or work in the fishing, lumbering, and mining industries, while many city dwellers have jobs in businesses and factories.

The Philippine Constitution guarantees freedom of worship. About 95 percent of all Filipinos are Christians. More than 80 percent of the population are Roman Catholics—a result of the Spanish colonization. The nation also has many Protestants, Muslims, and members of the Philippine Independent Church and the Philippine Church of Christ.

Most Filipinos have large families and maintain close relationships with family members, including elder relatives and distant cousins. Men traditionally hold positions of authority at home and in business. Today, however, many women work in professional fields, and growing numbers of women work in factories.

ETHNIC DIVERSITY

The people of the Philippines make up a tapestry of ethnic communities. Some of these groups live in rain forests or remote mountain provinces. Others live on the southern islands.

Northern cultures

Negritos, who arrived from Southeast Asia more than 30,000 years ago, were probably the first people to live in the Philippines. Today, Negritos live in small, isolated groups in various parts of the Philippines—in mountain jungles, along the shores of the northern provinces, and in the hills of Luzon. Negritos have traditionally lived by hunting and by gathering plants and fruits. Today, the Negritos face cultural extinction.

The Ifugao, a mountain people of northern Luzon, were the architects of a spectacular series of rice terraces. Two thousand years ago, the ancestors of the Ifugao carved terraces into the mountain slopes. The terraces, reinforced with stone or mud walls, provide flat fields on which to grow rice. The Ifugao also built an irrigation system to carry water from the high mountain forests to their fields. Generation after generation, the mountain people have maintained the terraced fields, which today are a UNESCO World Heritage site.

Muslim cultures of the south

In the 1300's, some 200 years before Magellan landed in the Philippines, Muslim traders made contact with the southern islands of Mindanao and the Sulu Archipelago. They brought the Islamic faith with them, and it took a strong hold among the local people.

When Spanish settlers arrived in the Philippines, they called these Muslim groups Moros. The name *Moros* is a form of the word *Moors*, the commonly-used name for the Arabic-speaking Muslims of North Africa, who had ruled parts of Spain from the 700's to 1400's.

A woman from northern Mindanao lives in an area with a mixed Muslim and Christian population.

Southwestern Mindanao is home to the T'boli people. Villagers, such as this family, draw water from a well and live in long houses set apart from one another on high ground or along ridges.

The Moros are composed of several culturally distinct groups. The Tausug, one of the largest groups and the first community to convert to Islam, were notorious pirates and smugglers. Today, the Tausug's long, narrow motorboats make them masters of the trade between islands.

The Maranao, another Muslim group, live in isolation in northern Mindanao. Their major town, Marawi, is a center of Islamic learning. It is located on Lake Lanao. The Maranao are known for their artistry and music. The *Darangen Epic* is an ancient song of about 72,000 lines that relates and celebrates the history and traditions of the Maranao people.

The Badjao, a third Moro group, are seafarers who have been born on their narrow-beamed wooden boats on the Sulu Sea. Some

Mountain people of northern Luzon live in small groups in isolated areas. One group, the Ifugao, are known for having built the magnificent rice terraces of Banawe.

Stilt villages, perched above the waters of the Sulu Sea in the southern Philippines, are home to the Badjao peoples, also known as "sea gypsies." Many of these people are born in their boats and spend their entire lives on the water.

of the Badjao, who are sometimes called "sea gypsies," spend their entire lives on the water, traveling to land only to die.

Several Muslim rebel groups in the Philippines have been fighting for independence since the 1970's. In 1996, the government signed a peace agreement with the largest rebel group, the Moro National Liberation Front. The agreement created a region of self-rule for Muslims in the southern Philippines.

Fighting continued between the government and two other Muslim rebel groups, the Moro Islamic Liberation Front (MILF) and the Abu Sayyaf Group (ASG). The MILF and the government signed cease-fire agreements in 2001 and again in 2003. However, the government refused to negotiate with the ASG because it used such tactics as kidnapping.

MANILA

The city of Manila became known as the *Pearl of the Orient* because of its beautiful setting and architecture. The city stretches along the east shore of Manila Bay on the island of Luzon. It covers a total area of about 15 square miles (38 square kilometers), and its metropolitan area covers 246 square miles (636 square kilometers).

The capital and largest city of the Philippines, Manila also has the country's leading port and serves as its major cultural, social, educational, and commercial center. Parts of the city are luxurious and wealthy, but others are areas of extreme poverty and slums. The Manila metropolitan area is one of the largest in the world.

When the Spanish founded Manila in 1571, they enclosed their city with high walls and a wide moat to protect themselves from unfriendly Filipinos. This area, called Intramuros, or the Walled City, stands on the banks of the Pasig River. The Pasig cuts through the heart of modern-day Manila and forms part of the harbor on Manila Bay.

The walls of Intramuros and some of the churches, convents, monasteries, and public buildings still stand, despite heavy bombing during World War II (1939-1945). Among the old buildings is the impressive San Agustin Church, completed in 1601.

Commercial hub

Manila is the banking, financial, and commercial center of the Philippines. It is also the headquarters of the Asian Development Bank, which lends money to promote economic growth in Asia. Manila has a wide variety of industries, including food processing, printing and publishing, shoe manufacturing, paint and varnish production, textiles, rope and cordage production, and soap, cigar, and cigarette manufacturing.

In addition, Manila's superb harbor and location make it an important port on Pacific and Far East trade routes. It is also a major air transportation hub.

A vendor sells fish in an open-air market on a Manila street. Seafood is a major part of the Filipino diet.

Government center

Spain surrendered Manila to the United States in 1898 after the Spanish-American War. The American administration installed a modern water-supply system and electric lighting. Japanese forces seized Manila in January 1942, and few buildings remained standing when the Japanese finally surrendered the city to U.S. troops in 1945. However, the Filipinos began rebuilding almost immediately.

Manila became the national capital of the Philippines when Independence was proclaimed on July 4, 1946. In 1948, the Philippine government made Quezon City the capital of the country. But Manila continued to serve as the seat of government, pending the completion of new government buildings in Quezon City. In 1976, the government again made Manila the country's official capital.

The crowded slums of Manila show the dark side of the capital. The city has a severe housing shortage. The Tondo district, along the waterfront, is the most densely populated section.

Manila stretches along the east shore of Manila Bay on the island of Luzon. The twisting Pasig River divides the city into two sections.

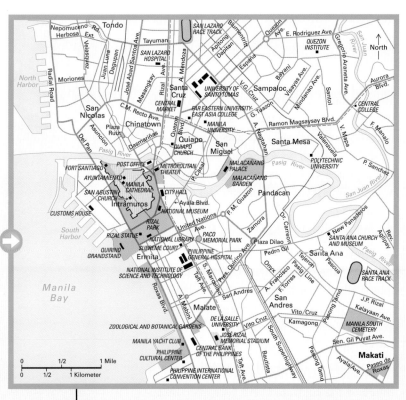

Cultural heart

Rizal Park, one of Manila's spacious parks, overlooks Manila Bay from just outside Intramuros. A scenic boulevard runs along the bay from the park, past the graceful mansions and lovely hotels of Manila. It also passes through the poor areas of the city.

A large Chinatown draws visitors on the north bank of the Pasig River. Many tourists travel to Quiapo, also north of the river, to visit its colorful market and lively restaurants, shopping centers, and movie theaters. Thousands of people go to the Quiapo Church every Friday to worship at the shrine of the miraculous image of the Black Nazarene, which is thought to have healing powers. This life-sized image of Christ bearing the cross was carved from black hardwood and brought from Mexico to Manila in the 1600's.

Manila has many universities. The oldest, the University of Santo Tomás, was founded in 1611. The city has a number of public libraries and museums, as well as a symphony orchestra and ballet and opera companies.

Makati is one of the cities that make up metropolitan Manila. It is one of the financial centers of Asia and ranks as the country's major financial, commercial, and economic center. Many international companies have their offices and headquarters in Makati.

ECONOMY

The islands of the Philippines are rich in natural resources. Mineral deposits, fertile farmland, lush rain forests, and coastal waters teeming with fish have traditionally been the cornerstones of the nation's economy. Today, agriculture, fishing, and forestry still employ about 40 percent of Philippine workers.

Agriculture and forestry

Filipinos farm only about 35 percent of the nation's land, but the farms produce most of the food needed for the entire population. Many farmers rent their land and pay the owner part of their crop.

Farmers grow rice and corn on about two-thirds of the cultivated land. Rice is an essential crop because most Filipinos eat it at every meal. Other leading food crops include sweet potatoes and *cassava,* a starchy root. Bananas, coconuts, mangoes, pineapples, and sugar cane are raised for both local use and export.

Philippine farmers also raise a crop called *abacá,* or Manila hemp, used in making rope. Mindanao, the island at the southeastern end of the Philippines, is one of the world's leading producers of abacá. The leaves of this plant contain a fiber that has great natural resistance to water, sun, and wind. Abacá fiber is also used in the manufacture of paper products. The plant has become less important on the world market, however, since the development of strong synthetic fibers.

More than 3,000 kinds of trees grow in the forests that cover about a third of the land of the Philippines. About 90 percent of the nation's timber comes from several related trees called *Philippine mahoganies.* Mangroves and pines also yield timber. Bamboo grows throughout the islands, and Filipinos use the stiff, hollow stems of this plant to build houses and to make baskets, furniture, and other items.

Fruits and vegetables make a tempting display in the market at Baguio in central Luzon. Fruit crops include bananas, coconuts, mangoes, and pineapples. Other leading crops include cassava, corn, rice, sugar cane, and sweet potatoes.

A farm girl works at a wet, muddy, but essential job—transplanting rice. The rice fields of central Luzon produce more of this important food crop than any other area of the Philippines.

Mining and fishing industries

Minerals make up a large part of the country's exports. Copper—found on Luzon, Cebu, Negros, and Samar—is the leading mineral. Large gold mines operate in northern Luzon. The Philippines also has deposits of chromite, coal, iron ore, limestone, manganese, nickel, silver, and zinc.

The waters surrounding the islands provide anchovies, mackerel, sardines, scad, tuna, and other fish. Divers near the southern islands gather sponges, clams, and oysters. Fish farmers raise milkfish, shrimp, and tilapia in ponds near the coast.

Business and industry

Service industries—including education, government, medicine, trade, transportation, communication, and financial services—employ about 45 percent of Philippine workers. Since the late 1900's, service industries have been the most rapidly expanding part of the economy.

Manufacturing, construction, and mining employ about 15 percent of the work force. The country's principal manufactured goods include cement, chemicals, cigars, clothing, foods and beverages, refined metals and petroleum,

The Pasig River connects Laguna de Bay to Manila Bay. The river stretches for about 15 miles (25 kilometers).

A miner pushes a cart loaded with gold ore inside a mine in Itagon, north of Manila. Such minerals as copper and gold make up a large part of the Philippines' exports.

Anchovies, one of the many types of fish found in Philippine coastal waters, are spread out to dry in the sun. Two other important fish, mackerel and tuna, are favorites in Philippine cooking.

sugar, textiles, and wood products. Many companies operate factories in *export-processing zones,* where businesses can import foreign goods without paying import taxes.

Contributing to the economic development in the Philippines is the nation's transportation system, one of the best in Asia. Despite the rugged terrain, roads cross the larger islands; railroads serve the largest island, Luzon; and a rapid transit rail system operates in the largest city, Manila. Ships and airplanes carry passengers and freight between islands. The leading ports are Manila, Cebu City, and Davao. Manila also has a major international airport.

The nation has a well-developed communications industry, with several television networks and a number of daily newspapers published in various languages.

HISTORY

In 1565, a group of Spanish explorers landed in the Philippines and claimed the islands for Spain. Under Spanish colonial rule, settlers divided the land among themselves, employing Filipinos as tenant farmers, laborers, and servants. Spanish priests converted most of the Filipinos to Roman Catholicism.

Spanish rule

Rebellion against Spanish rule arose in the late 1800's. The great opposition leader from that period was José Rizal, a doctor who worked for reform until 1896, when the Spanish executed him.

Emilio Aguinaldo also led a movement for independence during that period. In 1896, he took part in an unsuccessful revolt against Spanish rule. Spain promised political reforms if Aguinaldo would end the revolt and leave the islands. Aguinaldo then sailed to Hong Kong. However, the Spanish broke their promise, and Aguinaldo returned in May 1898. The United States gained possession of the Philippines that year as part of a peace treaty with Spain.

U.S. rule

Aguinaldo claimed that the United States had promised to make the Philippines independent immediately. He declared the establishment of the Philippine Republic on Jan. 23, 1899, and his troops began fighting the U.S. occupation forces. The Americans captured Aguinaldo in March 1901, however, and the fighting soon ended.

In 1901, the United States set up a colonial government in the Philippines. Under American rule, the English language spread rapidly. American businesses invested heavily in the Philippines, and the nation's economy became dependent on the United States.

The United States began to allow Filipinos to hold government positions. In 1935, the Philippines became a commonwealth with an elected government and a constitution similar to that of the United States.

On Dec. 7, 1941, Japanese planes bombed Pearl Harbor, a U.S. naval base in Hawaii, and the United States entered World War II (1939-1945) the next day. On December 10, Japanese

TIMELINE

c. 30,000 years ago	Negritos migrate to the Philippines.
c. 3,000 B.C.	Malays migrate from Malaysia and Indonesia.
1100's	Growth of Chinese influence.
1300's	Muslim traders introduce Islam.
1521	Magellan lands in the Philippines.
1565	Spanish explorers claim the Philippines for Spain and establish a permanent settlement.
1571	Spaniards begin construction of Intramuros (the Walled City) in Manila.
1600-1647	Raids by Dutch fleet from East Indies.
1600	First of several rebellions by Chinese population.
1762-1764	British occupy Manila during the Seven Years' War.
1896	The Spaniards execute José Rizal, a leader of the Philippine independence movement. Emilio Aguinaldo leads a revolt against Spain.
1898	Spanish-American War; U.S. gains possession of the Philippines.
1899	Emilio Aguinaldo declares the establishment of the Philippine Republic.
1901	U.S. subdues nationalist forces and sets up colonial government.
1901-1915	Rebellion of Muslim Moro peoples.
1935	U.S. grants commonwealth status to the Philippines.
1941	Japanese troops invade the Philippines.
1942-1944	Japanese occupy the country.
1944	MacArthur lands on Leyte with U.S. forces.
1946	The Republic of the Philippines established.
1949-1954	Philippine army defeats Communist-led Huk uprising.
1965	Ferdinand E. Marcos elected president.
1972	Marcos declares martial law.
1983	Benigno S. Aquino, Jr., assassinated.
1986	Protests force Marcos to resign; Corazon Aquino installed as president.
1989	Marcos dies.
1991	Mount Pinatubo volcano erupts. Clark Air Base, buried under ash, is closed.
1992	Fidel V. Ramos elected president. U.S. withdraws from Subic Bay Naval Base.
2000	President Joseph Estrada impeached on corruption charges.
2010	Benigno and Corzaon Aquino's son, Benigno Aquino III, is elected president.

Ferdinand Magellan
(1480?-1521)

Emilio Aguinaldo
(1869?-1964)

Corazon Aquino
(1933-2009)

troops invaded the Philippines. American and Philippine troops resisted the invasion until 1942, when most of the soldiers surrendered and were imprisoned. In October 1944, the United States sent fresh troops to the Philippines, defeating the Japanese several months later. The war hurt the Philippine economy and destroyed most of Manila.

Independence

The United States granted the Philippines complete independence on July 4, 1946. During the late 1940's, however, political problems and poverty caused widespread discontent among the Philippine people. The United States sent economic aid, and the economy of the Philippines began to improve as industries and trade increased.

In 1965, Ferdinand E. Marcos became president, leading the country into 20 years of repression and government corruption. In 1986, protests and unrest forced Marcos to hold a presidential election. His chief opponent was Corazon Aquino.

General Douglas MacArthur led U.S. troops as they liberated the Philippines from Japanese occupation during World War II. In this photo, MacArthur wades ashore on the island of Leyte.

Ferdinand Marcos's presidency (1965-1986) was marked by political repression and widespread corruption. When leading military officers staged a rebellion and thousands of people demonstrated against him in 1986, Marcos was forced to flee the country. After his death in 1989, his wife, Imelda (left), ran for president in 1992, but lost.

The legislature ruled that Marcos won the election, but many Filipinos accused his party of election fraud. Widespread protests broke out, and Marcos lost key military support. He then left the country, and Aquino was sworn in as president.

Aquino promised many changes for the Philippines, but she failed to carry out her policies, partly due to seven coup attempts against her government.

In 1998, Joseph Estrada was elected president, but in 2000, he was accused of corruption. The House of Representatives voted to impeach him, and the Senate took up the charges. Before a verdict was reached, Estrada lost the support of most politicians and stepped down. He was replaced by his vice president, Gloria Macapagal-Arroyo. In 2004, she won the presidential election in her own right.

In 2010, Senator Benigno Aquino III, son of Benigno and Corazon Aquino, was elected president.

Poland lies in central Europe and is bordered by the Czech Republic and Slovakia to the south and Germany to the west. To the north is the Baltic Sea, and to the east are Russia, Lithuania, Belarus, and Ukraine. The country's central location has made it a crossroads for invading armies throughout its history, thereby resulting in many boundary changes.

Poland is named for the Polanie, a Slavic tribe that lived more than 1,000 years ago in what is now Poland. The name *Polanie* comes from a Slavic word meaning *plain* or *field,* a fitting name for this land of flat plains, gently rolling hills, and clear blue lakes.

A country with a long and varied history, Poland once ruled an empire that stretched across much of central and eastern Europe. The Polish empire reached its height during the 1500's, achieving important advances in its cultural, economic, and political development. However, after a long period of decline, Poland ceased to exist as a separate state in 1795 and was divided between Russia, Prussia, and Austria.

An independent Poland came into existence again in 1918, after World War I (1914-1918). However, Germany and the Soviet Union divided the country in half barely 20 years later, at the beginning of World War II (1939-1945). Once again, Poland disappeared from the face of Europe until a new country was formed at the end of World War II.

The new Poland that was formed after the war was independent in name only. Its society and economy were entirely controlled by a Communist government that followed policies established by the Soviet Union. Not until the 1980's, when Poland was transformed from a one-party socialist system to a multiparty democracy, did the Poles regain a measure of their former freedom.

Even in the midst of continual political turmoil, the Poles have preserved their national identity, largely through their loyalty to the Roman Catholic Church. During the years between 1795 and 1918,

POLAND

when the Polish state did not exist, the people found themselves under pressure to become either "Germanized" or "Russianized." The Poles' Roman Catholic faith became the focus of their nationalism.

During the late 1940's and early 1950's, when Poland's Communist leaders tried to destroy the influence of the Catholic Church, it proved to be a powerful spiritual force that helped unify the people. Today, about 90 percent of the Polish people have been baptized in the Roman Catholic faith, and religious devotion goes hand in hand with patriotic fervor.

The Poles' national identity has also been preserved through the Polish language. During the 1800's, neither the Prussian nor the Russian authorities permitted the use of the Polish language in educational institutions. Despite these policies, the Polish language survived as a means of communication among the people. It was kept alive through songs, prayer books, and literary works.

Poland has produced many outstanding artists, musicians, and writers throughout its history. During the 1800's, when the Polish cultural identity was threatened by the Germans and the Russians, the paintings of Jan Matejko depicted scenes from Polish history, and the composer Frédéric Chopin wrote many works based on Polish dances. Leading Polish writers of the 1800's included the poet Adam Mickiewicz, the playwright Stanisław Wyspianski, and the novelist Henryk Sienkiewicz.

In the 1900's, many Poles won fame in the graphic arts, especially in poster design, which has long been recognized in Poland as a sophisticated art form. The Poles have also distinguished themselves in the cinematic arts. Polish film directors, such as Andrzej Wajda, Krzysztof Zanussi, and Krzysztof Kieslowski, have produced many compelling works.

73

POLAND TODAY

During the late 1940's, the Soviet Union gained increasing control over Poland, and from 1947 until 1989, the Polish government was directed by the Communist Party. Most Poles opposed Communist rule, but the Communists used police power and other methods to crush resistance. They took control of industry, forced farmers to give up their land and work on collective farms, and conducted fierce antireligion campaigns.

During the 1950's and 1960's, many Poles expressed their discontent through strikes and riots. Beginning in the mid-1970's, high prices and shortages of food and consumer goods triggered even more unrest. In the summer of 1980, as economic conditions worsened, workers in Gdansk and other cities went on strike, demanding better pay, free trade unions, and political reforms. In November of that year, the strikers won recognition for Solidarity, a free trade union headed by Lech Wałęsa. This was the first time a Communist country had recognized a labor organization that was independent of the Communist Party.

Meanwhile, the people continued to demand economic improvements and greater political freedom. In December 1981, General Wojciech Jaruzelski, the head of Poland's Communist Party, declared martial law and suspended Solidarity's activities. He also imprisoned Wałęsa and hundreds of other union leaders. Wałęsa and some union leaders were released in October 1982, and martial law formally ended in July 1983. The government continued to keep tight control over the people.

In the late 1980's, the tide began to turn. Soviet leader Mikhail Gorbachev made it clear that the Soviet Union would no longer enforce its will on its Eastern European satellite countries. The Polish government reached an agreement with Solidarity that led to the legalization of the union and to changes in the structure of the government. In 1989, a Senate was added to the Polish parliament, and the office of the president was created.

FACTS

Official name:	Rzeczpospolita Polska (Republic of Poland)
Capital:	Warsaw
Terrain:	Mostly flat plain; mountains along southern border
Area:	120,728 mi² (312,685 km²)
Climate:	Temperate with cold, cloudy, moderately severe winters with frequent precipitation; mild summers with frequent showers and thunderstorms
Main rivers:	Oder, Vistula, Warta
Highest elevation:	Rysy peak, 8,199 ft (2,499 m)
Lowest elevation:	Near Elblag, 7 ft (2 m) below sea level
Form of government:	Republic
Head of state:	President
Head of government:	Prime minister
Administrative areas:	16 wojewodztwa (provinces)
Legislature:	Zgromadzenie Narodowe (National Assembly) consisting of the Sejm with 460 members serving four-year terms and the Senat (Senate) with 100 members serving four-year terms
Court system:	Supreme Court
Armed forces:	121,800 troops
National holiday:	Constitution Day - May 3 (1791)
Estimated 2010 population:	38,025,000
Population density:	315 persons per mi² (122 per km²)
Population distribution:	61% urban, 39% rural
Life expectancy in years:	Male, 71; female, 80
Doctors per 1,000 people:	2.0
Birth rate per 1,000:	10
Death rate per 1,000:	10
Infant mortality:	6 deaths per 1,000 live births
Age structure:	0-14: 16%; 15-64: 71%; 65 and over: 13%
Internet users per 100 people:	44
Internet code:	.pl
Language spoken:	Polish
Religions:	Roman Catholic 89.8%, Eastern Orthodox 1.3%, Protestant 0.3%, other 8.6%
Currency:	Zloty
Gross domestic product (GDP) in 2008:	$526.97 billion U.S.
Real annual growth rate (2008):	4.8%
GDP per capita (2008):	$13,839 U.S.
Goods exported:	Cars, coal, food, iron and steel, machinery, petroleum products, ships
Goods imported:	Food, machinery, petroleum and petroleum products, pharmaceuticals, transportation equipment
Trading partners:	France, Germany, Italy, Russia, United Kingdom

Poland is a large nation in central Europe. It is bordered by the Czech Republic and Slovakia to the south and Germany to the west. To the north is the Baltic Sea, and to the east are Russia, Lithuania, Belarus, and Ukraine. The country's central location has contributed to its eventful history, during which the land has been invaded, conquered, and divided a number of times. Poland's present-day boundaries were established in 1945. Warsaw is the country's capital and largest city.

A Roman Catholic priest hears confession in Czestochowa, Poland's holiest shrine. The vast majority of Poles are Roman Catholic, and the Catholic Church—a strong opponent of Communism—played a key role in the political upheavals of the 1980's.

In the freest elections to take place in Poland in decades, non-Communist candidates were allowed to compete for all Senate seats and some seats in the lower house, the Sejm. Solidarity supporters took 99 of 100 Senate seats and gained a majority in the lower house. After the elections, the parliament elected Jaruzelski president, and Tadeusz Mazowiecki, a Solidarity leader, became the first non-Communist prime minister since World War II.

In 1990, the Communist Party was dissolved. Two social democratic parties were formed in its place.

Poland held its first direct presidential elections in December 1990. Lech Wałęsa won with more than 75 percent of the vote.

In January 1991, President Wałęsa selected Jan Krzysztof Bielecki as prime minister. Bielecki pledged to speed the pace of economic change in Poland, as the country faced the difficult transition from socialism to capitalism.

Poland held its first fully free parliamentary elections since World War II in October 1991. In the Senate, the Democratic Union, a party founded by former members of the Solidarity trade union, won the most seats. In the Sejm, 29 parties won representation.

Since the early 1990's, Poland's government has been led by a number of different parties. At first, these parties reflected tension over the change to a free market economy. At first, Solidarity and former Communist leaders held the presidency. In the early 2000's, major parties included Law and Justice, which promotes traditional social values, the pro-business Civil Platform, and the Democratic Left Alliance, which supports unions, equal rights, and socialist economic policies.

In 1999, Poland joined the North Atlantic Treaty Organization (NATO), a military alliance of Western nations. In 2004, the country joined the European Union (EU), an organization of European countries that cooperate in economics and politics.

HISTORY

The first rulers of what is now Poland were members of the Piast family, beginning with Prince Mieszko I, who controlled most of the land along the Vistula and Oder rivers by the 900's. His son, Bolesław I, conquered parts of what are now the Czech Republic, Slovakia, eastern Germany, and Ukraine. He became the first king of Poland in 1025. After his death later that year, the country experienced periods of warfare. The region eventually broke up into several sections, each ruled by a different noble.

A unified Poland

Poland was reunited in the early 1300's under Casimir the Great, the last Piast monarch. At the time of Casimir's death, Poland had a strong central government, a thriving economy, and a blossoming culture. In 1386, Queen Jadwiga of Poland married Władysław Jagiełło, the grand duke of Lithuania, and founded the Jagiellonian dynasty.

Poland prospered under the Jagiellonian dynasty and extended its boundaries to cover a large part of central and eastern Europe. After the mid-1500's, however, the monarchy began to lose power to the nobles, who dominated the national parliament. Poland lost much of its territory, and the decline of the empire continued into the 1700's.

In 1772, Austria, Prussia, and Russia took advantage of Poland's weakness and began to *partition* (divide) Polish territory among themselves. By 1795, after the third partition, the last remnants of Poland had disappeared.

In 1807, the French emperor Napoleon I gained control of Prussian Poland and made it into a Polish state called the Grand Duchy of Warsaw. However, after Napoleon's final defeat in 1815, Poland was again divided among Austria, Prussia, and Russia. A small, self-governing Kingdom of Poland was established under Russian control.

A new Polish state

After World War I (1914-1918), an independent Polish republic was formed. Under the Treaty of Versailles, Poland regained large amounts of territory from Germany, and the return of land in Pomerania, a region along the Baltic coast, gave Poland access to the sea. Poland's attempt

800's	Slavic tribes united under the Polanie.
966	Prince Mieszko I converts to Christianity.
1025	Bolesław I is crowned the first king of Poland.
Mid-1100's	Poland is divided into several sections.
1300's	Poland is reunified.
1333-1370	Casimir the Great rules Poland.
1386	Jagiellonian dynasty is founded.
1493	The first national parliament of Poland is established.
1500's	The Polish empire reaches the height of its powers.
1569	Poland and Lithuania are united under a single parliament.
1596	King Sigismund III moves the capital of Poland from Kraków to Warsaw.
1655	Poland loses most of its Baltic provinces to Sweden.
1772	Austria, Prussia, and Russia partition Poland.
1793	Prussia and Russia partition Poland.
1795	The third partition of Poland ends its existence as a separate state.
1849	Composer Frederick Chopin, whose works expressed Polish patriotism in the Polish harmonies and rhythms of his mazurkas and polonaises, dies in Paris.
1918	Poland becomes an independent republic.
1919	Poland regains territory from Germany under the Treaty of Versailles.
1919-1920	Poland enters war with Russia over partition land.
1926	Józef Piłsudski overthrows democratic government.
1939	Germany invades Poland; Germany and the Soviet Union partition the country.
1944	Polish Committee of National Liberation is formed in Lublin.
1945	A Communist-dominated government is formed; Poland's present-day boundaries are established.
1955	Poland signs the Warsaw Pact, a treaty that held most Eastern European nations in a military command under tight Soviet control.
1956	Workers in Poznan and other cities stage antigovernment riots.
1970	Strikes and riots break out in Gdansk and other cities.
1978	Polish Karol Cardinal Wojtyla becomes Pope John Paul II.
1980	The government recognizes Solidarity.
1981	General Wojciech Jaruzelski declares martial law and suspends Solidarity's activities.
1983	Solidarity leader Lech Wałęsa wins the Nobel Peace Prize.
1989	The government legalizes Solidarity; elections bring Solidarity-backed government to power.
1990	Poland's Communist Party is dissolved.
1991	Warsaw Pact is dissolved.
1997	A new constitution goes into effect.
1999	Poland becomes a member of NATO.
2004	Poland joins the European Union.
2010	President Lech Kaczynski and a number of government and military officials die in a plane crash in Russia.

Tadeusz Kosciuszko
(1746-1817)

Frédéric Chopin
(1810-1849)

Lech Kaczynski
(1949-2010)

Poland lost its independence in 1795, when its powerful neighbors divided it up among themselves. Poland reclaimed its independence in 1918, but came under Communist control after World War II.

Kraków, a city located in southern Poland, served as capital of the Polish kingdom from 1038 to 1596. This center of Polish culture is home to Jagiellonian University, founded in 1364.

Workers at the Lenin shipyard in Gdansk went on strike in 1980, demanding higher pay, free trade unions, and political reforms. The action forced the Communist government to recognize Solidarity, an organization of free trade unions.

to regain its territory to the east led to war with Russia in 1919 and 1920. The 1921 Treaty of Riga gave Poland some land on its eastern border.

Although the 1921 Constitution allowed for a democracy, the Polish government suffered from political instability. In 1926, the weak democratic government gave way to dictatorship under Józef Piłsudksi.

In August 1939, Germany and the Soviet Union signed a secret agreement to divide Poland among themselves. On September 1, Germany attacked Poland, and the Soviets invaded the country on September 17. The Poles were defeated within a month, after which Germany and the Soviet Union divided Poland. In 1941, Germany attacked the Soviet Union and seized all of Poland.

The rise of Communism

After the fall of Poland, a Polish government-in-exile was formed in Paris and later moved to London. In 1941, Polish Communists formed an exile center in the Soviet Union.

In 1944, the Soviet army invaded Poland and began to drive out the Germans. At the Yalta Conference in 1945, Allied leaders agreed to recognize the Polish Committee of National Liberation, which consisted almost entirely of Communists, as the provisional government of Poland if it was expanded to include representatives of the London government-in-exile. Agreements reached at the end of the war shifted Poland's borders westward, and the Soviet Union kept most of eastern Poland.

Most Poles opposed Communist rule. Nonetheless, the Polish Communists, with help from the Soviet Union, crushed resistance and established a Communist government by 1948.

ENVIRONMENT

Much of Poland is covered by the Great European Plain—a vast expanse of flat countryside that stretches across Europe from France to the Ural Mountains. Majestic, snow-capped mountains rise in southern Poland, but the country lacks natural barriers on the east and west, a factor that has contributed to centuries of invasion by its neighbors.

Outside the flat central plain, Poland's landscape becomes as varied and spectacular as any in Europe. Deep river gorges pierce the southern mountain ranges, while the Białowieska Forest—the last tract of primeval woodland in Europe—extends from the northeastern corner of Poland into Belarus. Fir, pine, oak, hornbeam, lime, ash, and other species of trees, some over 130 feet (40 meters) tall and more than 500 years old, provide shade for the European bison who roam there.

Land regions

Poland's seven land regions are the Coastal Lowlands, the Baltic Lakes Region, the Central Plains, the Polish Uplands, the Carpathian Forelands, the Sudeten Mountains, and the Western Carpathian Mountains.

The Coastal Lowlands extend in a narrow strip along the Baltic coast of northwestern Poland. Long stretches of sandy beaches and shifting dunes line much of the generally smooth coastline. About 12 square miles (32 square kilometers) of shifting sands in the village of Leba look so much like the deserts of North Africa that they were used in training troops for desert combat during World War II.

Most of northern Poland is covered by the Baltic Lakes Region, a hilly area dotted with small lakes. Forests and *peat bogs* (swamps made up of decayed plants) cover much of the land. Lumbering is the most important economic activity. In addition, some farmers raise wheat, sugar beets, potatoes, and rye. The beautiful, forested Masurian Lake District stretches from the Vistula River to Poland's eastern border. Poland's largest lakes, Lake Sniardwy and Lake Mamry, lie at the center of the region.

Sandy beaches line Poland's Baltic coast. Major coastal cities include Gdansk, Gdynia, and Szczecin. The Baltic coast has milder weather than the inland regions, while the mountain regions are cooler than the lowlands.

The town of Elk lies in northeastern Poland, a forested region with thousands of lakes. The lakes range in size from tiny woodland pools to large bodies of water. The lakes are surrounded by glacial debris deposited during the last ice age.

The soaring peaks of the Tatra Mountains, in southern Poland, are part of the Western Carpathian Mountains. Crystal-clear lakes, dark forests of pine and larch, and green meadows studded with rowan trees also characterize this region.

Canals connect these and many other small Masurian lakes. The region's scenic beauty makes it a popular vacation area for campers, hikers, and fishing enthusiasts.

South of the lake district and stretching across Poland, the Central Plains cover almost half the country. The plains are the country's major agricultural area and the site of some of Poland's most important cities, including Poznan, Warsaw, and Wrocław.

South of the Central Plains lie the Polish Uplands, a region of hills, low mountains, and plateaus. This densely populated area contains most of Poland's mineral wealth and much of its richest farmland.

South of the Polish Uplands, the land rises to the Carpathian Forelands, which lie within the branches of the Vistula and San rivers in southeastern Poland. With its fertile soil and important iron and steel industries, this region is one of the most densely populated areas in Poland.

Mountains and forests

The rounded peaks of the Sudeten Mountains—most of which are less than 5,000 feet (1,500 meters) high—border southwestern Poland. The Karkonosze Range, whose highest peak, Snieżka, rises to 5,256 feet (1,602 meters), lies in the central part of the Sudetens. To the east and toward the Oder River, the range loses altitude, and the taller peaks give way to rounded mountains covered by forests. The valleys and foothills of the Sudetens are used for agriculture.

In Poland's southeastern corner, the Carpathian Forelands rise to the Western Carpathian Mountains, which consist of the Beskid and Tatra chains. Dense forests cover the rounded domes of the Beskids, and only a few peaks rise above the tree line. The Tatra—a panorama of ancient craggy peaks, deep broad valleys, and numerous cascading waterfalls—rises beyond the Beskids. Rural towns and villages are nestled throughout the mountains in this part of Poland.

Small farming communities lie scattered across Poland's Central Plains, where farmers grow potatoes, rye, sugar beets, and other crops in spite of the poor, sandy soil. Near Warsaw, nature reserves protect elk, wolves, and wild boar.

WARSAW

Warsaw is the capital and largest city of Poland. It is also the nation's center of culture, science, and industry. It is situated in east-central Poland along both banks of the Vistula River. The city center and most of the residential areas are on the left, or west, bank of the river.

Early history

Warsaw traces its origins to a small Slavic settlement that existed as long ago as the 900's. In 1596, Warsaw became the capital of the Polish kingdom. Although Swedish invaders destroyed most of the city in 1656, Warsaw remained the capital until 1795, when Poland was divided between Austria, Prussia, and Russia.

Warsaw served as the capital of the Duchy of Warsaw, a state created by the French emperor Napoleon I, from 1807 to 1813. Russia then gained control of the city. Germany took over Warsaw in 1915, during World War I, and controlled it until Poland became independent in 1918.

The ravages of World War II

Warsaw was almost totally destroyed by the German army during World War II, and the city's people faced great suffering under the ruthless Nazi troops. The Germans took over the city in 1939, after a brutal, three-week siege.

During the German occupation, many of Warsaw's citizens were arrested and executed, and others were forced to leave the city. About 500,000 Jews were confined to a section called the *ghetto,* where many died of hunger and disease. Thousands of others were executed.

On Aug. 1, 1944, the people of Warsaw revolted against the Germans. Although the Soviet army had by that time reached the

The Column of King Sigismund—one of Warsaw's most famous landmarks—stands in Zamkowy Square, in front of the Royal Castle. Built in 1644, the monument honors the king who moved the capital from Kraków to Warsaw.

The towering Palace of Culture and Science, a gift from the Soviet Union in 1954, dwarfs the apartment and office blocks of modern Warsaw. The rapid growth of Warsaw's population after World War II resulted in severe housing shortages.

outskirts of the city, they did not come to the aid of the Poles, who were forced to surrender to the Germans on October 3.

To punish the Poles for their uprising, the Germans destroyed what was left of Warsaw. More than 200,000 residents of the city died in the uprising.

Rebuilding the city

Despite the destruction, old Warsaw lived on in the hearts of its citizens. In a remarkable labor of love, the people of Warsaw set about restoring their city soon after the destruction. They worked from the original plan of the city, which had been hidden from the Germans during the war.

Brick by brick, Polish architects, historians, builders, and masons painstakingly incorporated the remaining fragments of ancient buildings and monuments into authentic reconstructions. Often, they used old photographs—and even paintings and sketches—to re-create the city's architectural treasures.

The project was completed with the reconstruction of the Royal Castle, in the center of Old Town, in the 1980's. Today, with its narrow, winding streets and cobbled marketplace, Old Town provides a rare glimpse into Warsaw's past.

The Royal Castle is noted for the decorative paneling of its attic story. Similar attics, called *Warsaw lanterns,* crown many of the four- and five-story buildings surrounding Market Square *(Rynek).*

Other famous landmarks rebuilt since World War II include the Cathedral of St. John and the ancient city walls, both dating from the 1300's. Today, these buildings and monuments from Warsaw's past, risen from the ashes, stand amid modern hospitals, schools, and government buildings. They are a moving reminder of the spirit and determination of the Polish people, whose national anthem is entitled "Jeszcze Polska nie Zginęła" ("Poland Has Not Yet Perished").

Downtown Warsaw has wide streets lined with modern office buildings, apartments, and stores. The city is Poland's capital and a leading center of business and industry in eastern Europe.

The Golden Terraces is a commercial, office, and entertainment center in downtown Warsaw. The modern complex opened in 2007.

ECONOMY

When the Communists took control of Poland after World War II, the nation's economy depended largely on agriculture. The Communists, however, worked to create an industrial economy. New industrial regions were established around Kraków, Warsaw, and other cities.

Today, Poland is one of the leading industrial nations of Eastern Europe. The nation's factories manufacture chemicals, food products, iron and steel, machinery, ships, and textiles. With its rich coal fields in the south, Poland ranks among the leading coal-mining countries. It also produces large amounts of copper and silver.

Food subsidies and economic strain

In the early days of Communist rule, the government introduced massive subsidies on food in order to keep prices down and to win popular support for the regime. However, what began as a temporary measure was difficult to stop, and the government faced ever-increasing expenditures. Artificially low prices gave farmers no incentive to increase production, so the subsidies eventually led to food shortages.

In the early 1970's, the Polish government tried to solve its problems by borrowing money from countries in the West. The government used the money in an effort to modernize industries and to increase the production of consumer goods. The government planned to repay the loans with the profits from increased exports, but the expected boom in exports never took place.

Economic reform

When Poland's first non-Communist government since World War II took over in 1989, its new leaders prepared a radical plan to move the country from a centrally planned economy to a free market. The government encouraged the growth of private businesses, and it worked to transfer many industries from the state to cooperatives or private individuals. By shifting the emphasis from heavy industry to light industry, the government also hoped to create jobs and attract foreign investors with Poland's low labor costs.

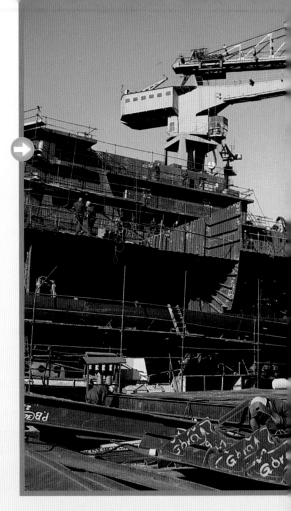

Shipbuilding, one of Poland's major industries, takes place in the ship-yards at Szczecin, on the Baltic Sea. Szczecin ranks as the nation's largest port. It also serves the Czech Republic, Hungary, and Slovakia.

Shoppers display garlic bulbs at a farmer's market in Zielona Góra. The city in eastern Poland is known for its historical churches and government buildings.

The owner of a clothing factory represents a new generation of Polish business people. Although government austerity programs jolted the economy in 1990, they also paved the way for private enterprise and a free market economy.

Polish miners in a pit near Katowice, in southern Poland, work one of the world's richest coal fields. Poland also has deposits of copper, lead, salt, silver, sulphur, and zinc.

People shop for produce and other agricultural goods in a market in Suwalki in the northwest corner of the country. Almost 20 percent of Poland's population works in agriculture.

The plan included the removal of price controls as well as most subsidies. To prevent inflation, strict wage controls remained in effect. Food became more widely available, but unemployment increased dramatically. The economy began to improve in the mid-1990's.

Business and production

Today, service industries—which produce services rather than goods—account for the majority of Poland's employment and of the value of all economic activity in the country. The government is the leading employer among Poland's service industries. It employs people in areas such as education, health care, and public administration. The second largest service industry employer is the trade, restaurant, and hotel sector of the economy. Tourism is important to these businesses. Millions of tourists visit Poland from the Czech Republic, Germany, and other countries every year. Other important service industry groups include finance, insurance, and real estate; and transportation and communication.

Manufacturing remains an important part of the economy. It employs about a fifth of the country's workers and contributes about a fifth of the value of all economic activity.

Agriculture employs nearly a fifth of the people, but it contributes less than 5 percent of the value of goods and services produced in Poland. Farmland covers most of the country, but the soil quality is poor. Poland is a leading producer of potatoes and rye. Other important crops include apples, barley, sugar beets, and wheat.

PORTUGAL

Comercio Square lies in the fashionable district known as the Baixa in Lisbon, the capital. In the center of the square stands a statue of King Jose I showing him on horseback. The district includes some of the city's finest shops.

Fishing villages line Portugal's coast, and fishing remains central to the way of life of many Portuguese people. About 40 percent of Portugal's people live in rural areas. Many of these people make their living on fish, and many others grow crops, including grapes that are used to make fine wine.

At the same time, Portugal's cities—especially Lisbon and Porto—are growing rapidly. Each year, a large number of the nation's rural people move to urban areas to find jobs in industry or other city activities. Portugal's cities have buildings that are hundreds of years old, as well as modern apartment and office buildings.

The Portuguese maintain close family ties. Often, two or more generations of a family live together in the same house. Men and women who move to cities from villages tend to keep in close touch with their relatives back home.

In the 1400's and 1500's, Portugal was one of the mightiest nations in the world. The vast Portuguese Empire once ruled colonies in Africa, Asia, and South America. The discoveries of such famous Portuguese explorers as Bartolomeu Dias, Vasco da Gama, and Pedro Álvares Cabral began the Great Age of European Discovery.

During Portugal's Golden Age, from the 1400's to the 1600's, the arts flourished. Painters and architects celebrated their country's achievements in their artistic masterpieces. All too soon, though, the Portuguese Empire declined. Held back by weak leadership, unstable governments, and a poor economy, Portugal has struggled to overcome its problems ever since.

Today, Portugal is one of the poorer countries in Europe. The country has generally experienced economic growth since the 1960's, but fighting between political groups, along with periods of inflation and unemployment, have slowed progress.

Nevertheless, Portugal remains a beautiful country. Its natural charm is exemplified in its unspoiled coastline, dotted with quaint fishing villages. Enhancing Portugal's beauty is the country's rich cultural heritage, which comes vividly to life in its folk music and dance, religious festivals, cuisine, and delightful customs.

PORTUGAL TODAY

Modern Portugal came into being in 1974, when military leaders overthrew a dictatorship that had ruled the country since 1928. The revolution, known as the Armed Forces Movement, restored the rights of the people, abolished the secret police, and allowed the formation of political parties. Clashes between the new political parties led to further violence in 1974 and 1975. But for the most part, Portugal entered its new period of democracy with little trouble.

A new government

In 1976, Portugal adopted a new constitution that guarantees such rights as freedom of speech, freedom of religion, and freedom of the press. Portugal is now a republic, with a parliament and a president elected by the voters. The prime minister is usually the head of the majority party in the parliament, which is known as the Assembly of the Republic.

Control of Portugal's government has changed hands many times since 1976. The main political parties have often banded together in *coalitions* that have worked together to control the government. However, the lack of cooperation between the political parties has been a continuing problem. Portugal remains one of the poorest nations in Europe. Periods of economic growth have been hampered by high unemployment, excessive government debt, and inflation.

In 1985, the Social Democratic Party won a majority in the parliament without having to share power in a coalition. It controlled the government for the next 10 years and brought some political stability to the country. In 1995, Portugal's Socialist Party gained the largest number of seats in the Assembly. The Social Democrats won the most seats in the Assembly in 2002, but the Socialist Party regained control in 2005. Socialist Prime Minister José Sócrates resigned during an economic crisis in 2011. The Social Democrats won a parliamentary election and formed a coalition government.

FACTS

Official name:	Republica Portuguesa (Portuguese Republic)
Capital:	Lisbon
Terrain:	Mountainous north of the Tagus River, rolling plains in south
Area:	34,350 mi^2 (88,967 km^2)
Climate:	Maritime temperate; cool and rainy in north, warmer and drier in south
Main rivers:	Tagus, Douro, Guadiana, Mondego, Sado
Highest elevation:	Estrela, in Serra da Estrela, 6,539 ft (1,993 m)
Lowest elevation:	Atlantic Ocean, sea level
Form of government:	Republic
Head of state:	President
Head of government:	Prime minister
Administrative areas:	18 distritos (districts), 2 regioes autonomas (autonomous regions)
Legislature:	Assembleia da Republica (Assembly of the Republic) with 230 members serving four-year terms
Court system:	Supremo Tribunal de Justica (Supreme Court)
Armed forces:	42,900 troops
National holiday:	Portugal Day - June 10 (1580)
Estimated 2010 population:	10,221,000
Population density:	298 persons per mi^2 (115 per km^2)
Population distribution:	58% urban, 42% rural
Life expectancy in years:	Male, 75; female, 82
Doctors per 1,000 people:	3.4
Birth rate per 1,000:	10
Death rate per 1,000:	10
Infant mortality:	4 deaths per 1,000 live births
Age structure:	0-14: 16%; 15-64: 67%; 65 and over: 17%
Internet users per 100 people:	42
Internet code:	.pt
Languages spoken:	Portuguese (official), Mirandese
Religions:	Roman Catholic 84.5%, other Christian 2.2%, other 13.3%
Currency:	Euro
Gross domestic product (GDP) in 2008:	$244.49 billion U.S.
Real annual growth rate (2008):	0.2%
GDP per capita (2008):	$22,897 U.S.
Goods exported:	Clothing and footwear, food and wine, machinery, motor vehicles, petroleum products
Goods imported:	Crude oil and petroleum products, fish, food, iron and steel, machinery, pharmaceuticals, transportation equipment
Trading partners:	Belgium, France, Germany, Italy, Netherlands, Spain, United Kingdom, United States

The Amoreiras shopping and office complex adds drama to the skyline of Lisbon. The city, which also serves as Portugal's economic and cultural center, lies at the mouth of the Tagus River near Portugal's southwest coast.

War in the colonies

During the 1960's, Portugal's economy was seriously weakened by costly wars in the nation's African colonies. The wars started when rebels in the colonies of Angola, Mozambique, and Portuguese Guinea began armed resistance against their governments. Portugal sent troops to stop the rebellion, and thousands of people were killed on both sides of the struggle.

After the 1974 revolution, the provisional government ended Portuguese rule in the African colonies. In 1974, Portuguese Guinea was the first colony to gain its independence, becoming the nation of Guinea-Bissau. A year later, Angola, Cape Verde, Mozambique, and São Tomé and Príncipe became independent. In 1976, Indonesia took over Portugal's colony of Portuguese Timor in the East Indies.

Portugal now rules only its mainland territory and the Azores and the Madeira Islands. Portugal had control of the tiny territory of Macao on China's southern coast until 1999, when Macao passed back to Chinese control.

Looking to the future

Until the mid-1900's, Portugal's economy was based primarily on agriculture and fishing. Today, manufacturing is a growing element in the diversification of the economy, accounting for about 30 percent of Portugal's economic production. Service industries, taken together, account for about 60 percent of the economic production.

In 1986, Portugal joined the European Community, an economic organization that later became the European Union (EU). By participating in the EU, Portugal's leaders hope to promote further economic growth and raise the country's standard of living.

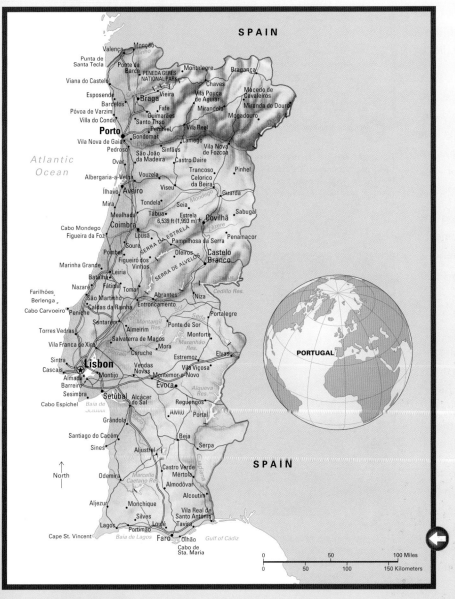

The nation of Portugal lies at the western edge of continental Europe. In addition to its mainland territory, which shares the Iberian Peninsula with Spain, Portugal includes the Azores and the Madeiras, two island groups in the Atlantic Ocean.

ENVIRONMENT

Portugal is a small, narrow country situated on the western edge of the Iberian Peninsula. It extends 350 miles (563 kilometers) from north to south, but only 125 miles (201 kilometers) from east to west at its widest point. The Serra da Estrela mountain range, which cuts across the central part of the country, serves as a dividing line between northern Portugal and southern Portugal.

Coastal Plains

Portugal's broad Coastal Plains extend along the country's west coast. Numerous lagoons have formed where land meets water as the plains slope gently to the sea. Farmers grow rice in the flat river valleys of the plains, and corn is cultivated in the higher, drier areas. Lines of trees protect the fields against erosion by the wind.

In the southern Coastal Plains, large villages are set amid olive groves, vineyards, and cornfields. A wide depression near the mouth of the Tagus River has collected sediments from the sea and now provides very fertile land, yielding large crops of olives, rice, and wine grapes.

However, the land in this area is still sinking and is geologically unstable—a condition that sometimes causes earthquakes. Portugal's most serious earthquake occurred in 1755 and practically destroyed the city of Lisbon.

Northern Portugal

East of the Coastal Plains, northern Portugal consists of the mountains and plains of the Northern Tablelands. This region begins at the Minho River, which separates Portugal and Spain to the north, and extends to the Central Range, south of the Mondego River.

A continuation of Spain's Meseta, the Northern Tablelands consist mainly of plains broken up by mountain ranges. The Douro River cuts deeply into the land, forming a narrow, sunny valley. Grapes used in the production of Portugal's famous port wines are cultivated on steep slate terraces in the Douro Valley.

Portugal can be divided into four main land regions: the Coastal Plains, the Northern Tablelands, the Central Range, and the Southern Tablelands.

The pine forests around Coimbra in northern Portugal yield resin, used in making turpentine, and lumber, used in making furniture. Cork bark is collected from the cork oak trees that grow in the southern and central regions.

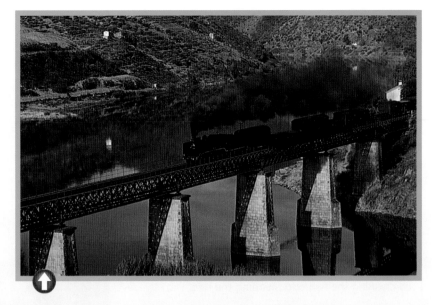

A locomotive steams across the Douro River in northern Portugal. The Douro meets the Atlantic Ocean at the city of Porto. The river provides hydroelectric power, and the surrounding valley yields grapes for making port wine.

Large farms surrounded by vast, rolling plains make up much of southern Portugal's landscape. Although southern Portugal has a number of larger, state-owned collective farms, most crop farms are small, family-owned operations.

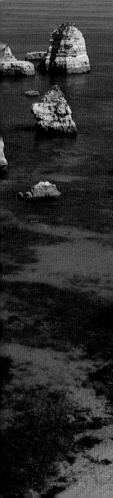

Bright sunshine, clear waters, and sandy beaches make the southern coast of Portugal a popular vacation spot. Temperatures are mild throughout the country, ranging from about 70° F (21° C) in July to about 50° F (10° C) in January.

In the northeast, the high plateaus of the Northern Tablelands rise to heights of 1,640 to 2,460 feet (500 to 750 meters). The plateaus are barren, desolate areas, with little vegetation. Closer to the west coast, small villages fringed by forests of beech and oak nestle in the meadows and fields.

South of the Douro River, the Northern Tablelands slope down to the plateau of Beira, a hilly region rising between 2,000 and 3,300 feet (600 and 1,000 meters). Beira is covered with olive groves and the Mediterranean shrub known as *maquis*.

Southern Portugal

The Serra da Estrela, the dividing line between northern and southern Portugal, is a region of bare granite uplands, formed during the Ice Age. The peaks of the Serra da Estrela, which rise more than 6,000 feet (1,829 meters) above sea level, are snow-covered for several months each year.

South of the Serra da Estrela, the landscape consists of a series of wide plateaus covered by miles of wheat fields and pastureland. This gently hilly country develops into wide fields and bogs near the coast.

Farther south, the low plateaus merge with the barren, rounded hills of the High Algarve, a highland area. Farms in the Lower Algarve produce crops of almonds, carobs, figs, olives, and oranges, and cork oak trees.

PEOPLE

Portugal has a population of about 10 million. About 40 percent of the country's people live in rural areas. About 60 percent live in cities and towns.

Most rural Portuguese live in small fishing or farm villages. Fishing villages line the country's coast. The people of these settlements have long relied on fishing for their livelihood. Many people brave the rugged waters of the Atlantic Ocean in small boats to catch fish. Others do such chores as cleaning the fish and mending the nets.

Portuguese farmers raise a variety of crops, but they are best known for their fine grapes that are used to make wine. Wines from Portugal are enjoyed by people in many parts of the world.

Life in the capital

The capital and largest city of Portugal, Lisbon is the nerve center of the nation. It boasts one of Europe's most important natural harbors. Large shipments of Portuguese ceramics, cork, sardines, and tomato paste enter and leave the port of Lisbon every day. Just across the Tagus, one of Europe's major shipyards bustles with activity.

Lisbon has a long history, beginning more than 3,000 years ago. Ancient Greeks, Carthaginians, and Romans colonized Lisbon before the Moors seized it during the A.D. 700's. Later, in the 1100's, Christian forces led by King Afonso I reclaimed Lisbon. The city became the official capital of Portugal in the late 1200's. It was in Lisbon that the great Portuguese explorers and adventurers planned their expeditions to the New World.

However, little remains of these early days, because most of the city was destroyed on the morning of Nov. 1, 1755, by one of the worst earthquakes in history. Within a few hours after the quake, huge tidal waves swept up the Tagus River from the Atlantic, flooding the city and drowning many of its inhabitants. Many more people were killed by the raging fires that fol-

Tourists shop along Rua Augusta, a pedestrian zone in Lisbon's Baixa district. On the north side of the street is a triumphal arch that frames the statue of King Jose I in Comercio Square.

A fisherman at Nazaré, on Portugal's central Atlantic coast, repairs his nets. Traditionally a seafaring people, the Portuguese still brave the rough waters of the Atlantic Ocean in small boats. The catch includes cod, sardines, tuna, and many other kinds of fish.

lowed. By the time it was all over, more than 60,000 people had died and about two-thirds of Lisbon lay in ruins.

Portuguese architects soon planned a new city on top of the ruins. They introduced the wide, graceful boulevards, mosaic pavements, and symmetrical buildings that now give Lisbon its elegant character. Sadly, many of the buildings so masterfully constructed during this period were destroyed by fire during the summer of 1988.

Today, the city is a modern metropolis, complete with skyscrapers, an international airport, and a subway system. The people live in pleasant, pastel-colored houses and apartment buildings. Yet—even now—Lisbon retains an Old World quality in the cobbled streets and alleyways of *Alfama,* its oldest quarter.

Workers gather to relax in Alfama after a long day's work. Formerly a Moorish town, Alfama is Lisbon's oldest quarter. Alfama is an Arabic word, as are many other words in the Portuguese language. Alfama's steep, narrow streets preserve the unique charm of the city's past.

People bask in the sun in one of the popular cafes along the Rua Augusta. The city's shops and restaurants benefit from the large number of tourists that visit the country.

Northern city dwellers

North of Lisbon in the central part of the country, the city of Coimbra is the site of the University of Coimbra—one of the oldest universities in Europe. Along with its academic atmosphere, Coimbra keeps alive the folk tradition of *fado,* a popular form of Portuguese song accompanied by guitar.

Farther north, on the banks of the Douro River, lies Porto, Portugal's second-largest city. From its ancient beginnings as a Roman trading community, Porto has grown to become the commercial and industrial center of northern Portugal. The city is also known for its role in processing and exporting Portugal's famous port wines.

About 50 miles (80 kilometers) north of Porto lies the city of Braga, often called *Portuguese Rome.* Braga's Holy Week procession is considered to be the finest in all of Portugal.

ECONOMY

Although its economy has long been based on agriculture and fishing, Portugal has recently experienced a growth in industrialization. Manufacturing now accounts for about 30 percent of the country's economic production, whereas agriculture and farming together account for 10 percent. Service industries make up the remaining 60 percent.

Portugal's 1986 entry into the European Community, now the European Union (EU), provided a welcome boost to the nation's economy. In 1999, Portugal joined the EU's common currency, the euro. However, during the first decade of the 2000's, rising unemployment and growing government debt forced the government to implement harsh austerity measures and request a loan from the EU.

Natural resources

Portugal has valuable natural resources, but for the most part, they remain undeveloped. Building stone, found throughout the country, has become Portugal's most important developed mineral resource.

The mountains of Portugal hold deposits of coal, copper, and wolframite, which is used in making tungsten. However, mining has not been well developed in Portugal.

Water is also one of Portugal's natural resources. The country's rivers, particularly the Tagus and the Douro, provide valuable hydroelectric power for Portuguese factories and homes.

About a third of Portugal is covered by forests. Pine trees grow in northern Portugal, while cork oak trees thrive in the central and southern regions of the country. Cork is collected from the cork oak trees for use in making such products as bottle stoppers, bulletin boards, fishing floats, golf balls, and life jackets.

Industry and agriculture

The leading manufacturing category of Portugal is textiles, with cotton fabric being the most important textile produced. Portuguese factories also produce cement, ceramics, cork products, shoes, and fertilizer.

Portugal's farmers raise a variety of crops, including almonds, corn, olives, potatoes, rice, tomatoes, and wheat. Farmers also raise cattle, chickens, hogs, and sheep. The most important agricultural product, however, is wine grapes.

The vineyards of the Douro Valley produce the grapes used in making port wine. Port is a *fortified wine* popular all over the world. Fortified wines have brandy or wine alcohol added to them, and they tend to be sweeter than most other wines. Portugal also makes a variety of red and white wines.

Portuguese tomatoes, noted for their delicious flavor, are sorted at a processing plant. In French cooking, dishes made with tomato sauce are often described as à la Portugaise (in the Portuguese style).

Barrels of wine are transported down the Douro River in picturesque boats displaying the names of wine merchants on their sails. These vessels are among the few surviving examples of barcos rabelos, the flat-bottomed boats used to carry port wine from the vineyards to the city of Porto for export. Every year on June 24, during the feast of São Joao (Saint John), the boats sail in a race on the river.

THE CORK OAK TREE

Cork comes from the cork oak tree, which thrives in warm, sunny climates. Cork is a lightweight, spongy substance obtained from the bark of the tree. A cork tree must be about 20 years old before its bark is thick enough to be stripped, and then it can be stripped at intervals of about 8 to 10 years. The first layer removed is called virgin bark. Workers use a long-handled hatchet to cut long, oblong sections of bark from the top of the lowest branches to the bottom of the tree. These sections are boiled, scraped, straightened, and then dried in the sun. The cork is packed in bundles and graded for quality and thickness.

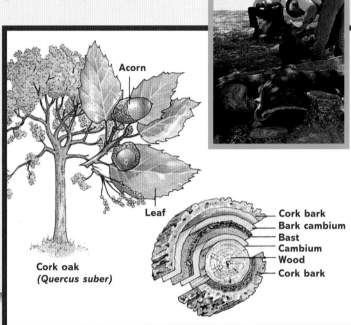

Acorn

Leaf

Cork oak
(Quercus suber)

Cork bark
Bark cambium
Bast
Cambium
Wood
Cork bark

Portuguese farms are generally quite small, and the farmers who own them often use old-fashioned methods and equipment. As a result, the farms are not as efficient as they could be, and production may be low. Since the 1974 revolution, the Portuguese government has set up large collective farms, mostly in the southern region of the country. These state-owned farms use up-to-date methods and modern equipment, which increase productivity. Properties are gradually being returned to private owners.

Fishing

Every day, Portuguese fishing crews set sail on the Bay of Biscay—and even far into the Atlantic Ocean—in search of deep-sea catches. Hundreds of deep-sea trawlers bring ashore tons of fish and shellfish each year. Cod, sardines, tuna, and other kinds of fish make up the bulk of the haul.

Fishing crews check their nets after a day at sea. Many still use the traditional fishing boats of their ancestors. These boats are flat-bottomed, rising to a high, pointed bow and stern. Oxen are used to drag them out of the water, out of range of the pounding surf.

HISTORY

The recorded history of Portugal begins about 5,000 years ago, when a tribe called the Iberians occupied the area. Later, the Phoenicians, the Celts, the Greeks, and the Carthaginians invaded the peninsula.

After the Romans defeated the Carthaginians in 201 B.C., they claimed rights to the entire Iberian Peninsula, but it took them about 200 years to complete their conquest. The Romans named the Portuguese portion of their empire *Lusitania*. They called the city of Porto *Portus Cale*—the origin of Portugal's present name.

The Romans built cities and linked them with their system of roads. The use of Latin spread through the region, and Latin became the root of both the Portuguese and Spanish languages.

During the A.D. 400's, Germanic tribes swept across the Pyrenees to take over the country. In 711, Muslim armies from North Africa invaded the peninsula. They brought advanced concepts of architecture, improved roads and education, and new crops and farming methods, such as irrigation in the dry south.

By the mid-1000's, the Christian forces of Iberia were gaining strength in their battle to drive out the Muslims. Meanwhile, King Alfonso VI, ruler of the Spanish Kingdom of Castile, named Henry of Burgundy the Count of Portugal. Henry was a French nobleman who had joined the Christians in their struggle against the Muslims.

Henry's son and successor, Afonso Henriques, won many battles against the Muslims. Afonso Henriques also broke away from Castile in 1143 and declared himself king of the independent kingdom of Portugal. By the mid-1200's, the Muslims were driven from Portugal. In 1297, Castile officially recognized Portugal's borders, which have remained almost unchanged ever since.

In 1385, Portugal and England signed the Treaty of Windsor. This political alliance is the oldest in Europe still in force.

c. 3000 B.C.	Iberian tribes inhabit what is now Portugal.
1000's B.C.	Phoenicians from the eastern shore of the Mediterranean establish settlements on the Iberian Peninsula.
400's B.C.	Carthaginians control much of the Iberian Peninsula.
201 B.C.	Romans defeat Carthaginians and begin conquest and occupation of Iberian Peninsula.
c. 19 B.C.	Rebellious local tribes finally brought under Roman rule.
A.D. 400's	German tribes take over country.
711	The invasion of the Moors begins.
1096	Henry of Burgundy becomes Count of Portugal.
1143	Afonso Henriques establishes Portugal as an independent kingdom.
1212	Battle of Las Navas de Tolosa breaks Moorish power.
1297	Castile recognizes the boundaries of the Portuguese kingdom.
1385	Portugal enters into a political alliance with England.
1419	Prince Henry the Navigator launches the first Portuguese expedition to the West African coast.
1494	Treaty of Tordesillas divides the New World between Portugal and Spain.
1497-1498	Vasco da Gama sails around the Cape of Good Hope and on to India.
1500	Pedro Álvares Cabral claims Brazil for Portugal.
1580	Spain invades and conquers Portugal.
1640	Rebellion restores Portugal's independence.
1807	French forces under Napoleon I invade Portugal.
1822	Portugal loses its colony of Brazil.
1908	King Carlos I is assassinated.
1910	The monarchy is overthrown, and Portugal becomes a republic.
1914-1918	Portugal fights in World War I (1914-1918) on the side of the Allies.
1926	Army officers overthrow Portugal's civilian government.
1932-1968	António de Oliveira Salazar rules as dictator.
1960's	African colonies revolt against Portuguese rule.
1974	Military officers overthrow the dictatorship.
1975	Almost all remaining Portuguese colonies are granted independence.
1976	First free elections in more than 50 years.
1986	Portugal joins the European Community (now the European Union (EU).
1999	Macao is returned to Chinese rule.
2011	Prime minister resigns amid a severe economic crisis in which Portugal is forced to request a bailout from the EU.

Prince Henry the Navigator (1394-1460) encouraged exploration of West Africa.

The Marquis of Pombal (1699-1781) was prime minister of Portugal under King José I.

Mário Soares (1927-) opposed António de Oliveira Salazar's dictatorship.

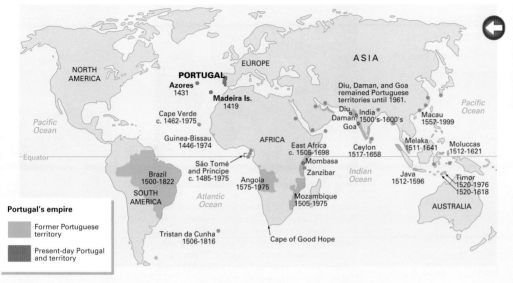

NORTH AMERICA

EUROPE

ASIA

PORTUGAL
Azores
1431

Madeira Is.
1419

Cape Verde
c. 1462-1975

Diu, Daman, and Goa
remained Portuguese
territories until 1961.

Diu
Daman
Goa

India
1500's-1600's

Macau
1557-1999

Pacific
Ocean

Pacific
Ocean

Guinea-Bissau
1446-1974

AFRICA

East Africa
c. 1505-1698

Ceylon
1517-1658

Melaka
1511-1641

Moluccas
1512-1621

Equator

São Tomé
and Príncipe
c. 1485-1975

Mombasa

Zanzibar

Indian
Ocean

Java
1512-1596

Timor
1520-1976
1520-1618

Brazil
1500-1822

Angola
1575-1975

SOUTH
AMERICA

Atlantic
Ocean

Mozambique
1505-1975

AUSTRALIA

Portugal's empire

Former Portuguese
territory

Present-day Portugal
and territory

Tristan da Cunha
1506-1816

Cape of Good Hope

Portugal's empire in the mid-1500's, at the height of the nation's power, included colonies in many parts of the world. This map shows these colonial possessions and when Portugal ruled them.

The age of expansion

The Portuguese had long been a seafaring people. They were excellent sailors and knew a great deal about navigation and shipbuilding. Under the patronage of Prince Henry the Navigator, Portuguese sailors developed sailing skills that later helped Portuguese explorers span the globe and build a powerful empire.

Although Prince Henry never went on a sea voyage himself, he was very skillful at helping mapmakers, navigators, and explorers work together. He also built a famous navigation school at Sagres.

Prince Henry encouraged and sponsored many expeditions. Through his efforts, Portuguese explorers reached the Madeira Islands and the Azores. By the time Prince Henry died in 1460, the explorers had sailed along the west coast of Africa to Sierra Leone.

In 1497, Vasco da Gama took four ships on a daring voyage around Africa's Cape of Good Hope. By 1500, the Portuguese had reached the coasts of Africa, the Arabian and Malay peninsulas, the East Indies, and the Orient.

Years of decline

At its height, Portugal's empire stretched from Brazil to China. Resources from its colonies—including gold from Africa and diamonds from Brazil—brought great wealth to Portugal. However, the small country found it too difficult to hold and manage such a vast empire. During the 1600's, England, the Netherlands, and France began to take over parts of the empire.

In 1580, Spain invaded Portugal and occupied the nation for 60 years, until Portuguese independence was restored in 1640. In 1807, Portugal was again invaded and conquered, this time by Napoleon's French armies. From 1808 to 1811, British forces helped the Portuguese drive out the French.

By the time Portugal's King John VI returned to the throne after the French occupation, a new political spirit had grown strong in Europe. Many Portuguese began demanding greater representation in their government, but for many years, little progress was made toward a republican government. Finally, in 1910, the monarchy was overthrown, and Portugal became a republic.

During the next 15 years, Portugal suffered through the political instability created by 45 different governments. In 1926, army officers overthrew the civilian governments. Dictators—most notably António de Oliveira Salazar, prime minister from 1932 to 1968—ruled the country until Portugal adopted a democratic system of government in 1976.

PUERTO RICO

Puerto Rico, a commonwealth of the United States, is a beautiful tropical island of the West Indies, about 1,000 miles (1,600 kilometers) southeast of Florida. The U.S. Congress is responsible for governing Puerto Rico, but the island has wide powers of self-rule.

Most U.S. federal laws apply to Puerto Rico as if it were a state. Voters elect a governor, who serves as the chief executive. The people also elect members of their two-house legislature.

The people

The people of Puerto Rico are U.S. citizens and can move to the mainland without immigration restrictions. However, when they live in Puerto Rico, they cannot vote in U.S. presidential elections, and they do not pay federal income taxes.

Puerto Rico is a crowded island with a population of about 4 million. Most of the people live in urban areas. San Juan, the capital and largest city, has a metropolitan area of more than 2-1/2 million people.

Most Puerto Ricans are of Spanish descent and speak Spanish, though many also speak English. There are smaller numbers of people of Portuguese, Italian, French, and African backgrounds. Arawak Indians lived on the island before it was colonized by Spain, but nearly all were killed or died of disease. No full-blooded Indians remain on Puerto Rico, but some people have mixed Indian and Spanish ancestry.

The economy

Millions of tourists visit Puerto Rico every year. They are attracted by its balmy climate. Average temperatures range from 73° to 80° F (23° to 27° C), and sea breezes make the temperature especially comfortable.

The climate attracts not only tourists but also businesses. Manufacturing is the single most valuable industry. Many Puerto Ricans work in factories that produce chemicals, clothing, computer and electronic products, electrical equipment, food products, machinery, medical equipment and supplies, petroleum and coal products, rubber and plastic products, and transportation equipment. Many factories were set up under

Coconut palms rise from the sand on the Puerto Rican coast. Tourists—mostly from the United States—are attracted by these beautiful beaches and the balmy climate. Tourists spend hundreds of millions of dollars every year in Puerto Rico's shops, hotels, and restaurants.

Operation Bootstrap, a government program that helped businesses find locations, finance construction, and train workers.

The climate also allows Puerto Ricans to grow tropical crops. In addition to sugar cane and coffee—the leading crops—farmers grow avocados, bananas, citrus fruits, coconuts, pineapples, plantains, and tobacco. Milk, poultry, and eggs are

Rincón, on Puerto Rico's northwest coast, is a popular destination for visitors who enjoy the resorts and beaches on the Caribbean Sea.

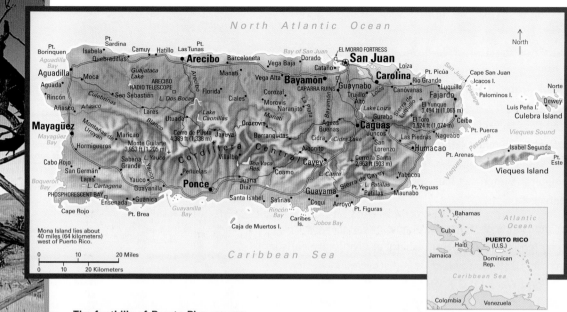

North Atlantic Ocean

Mona Island lies about 40 miles (64 kilometers) west of Puerto Rico.

0   10   20 Miles

0   10   20 Kilometers

Caribbean Sea

Bahamas

Atlantic Ocean

Cuba

PUERTO RICO
(U.S.)

Haiti

Jamaica

Dominican Rep.

Caribbean Sea

Colombia   Venezuela

The foothills of Puerto Rico are carpeted with dry grass. The Cordillera Central (Central Mountain Range) runs east and west across the island.

Puerto Rico has four main land regions. They are the Coastal Lowlands, the Coastal Valleys, the Foothills, and the Central Mountains.

El Morro Fortress commands the coast near San Juan. Built by Spain in 1539, the fort is a reminder of Puerto Rico's colonial past.

Spaniards, led by Juan Ponce de León, established the first European settlement in 1508. For more than 300 years, the island colony grew slowly, through hurricanes, plagues, and attacks by the Carib Indians and the Dutch, English, and French. By the late 1800's, Puerto Ricans had been given a large amount of local rule by Spain.

When the Spanish-American War began in 1898, U.S. forces landed in Puerto Rico. Spain surrendered the island to the United States when it lost the war.

A civil government was established for Puerto Rico in 1900, and in 1917 Puerto Ricans were made U.S. citizens. In the early 1940's, Operation Bootstrap began to improve living conditions on the island. In 1947, Puerto Ricans elected their first governor. The Puerto Rican people and Congress approved a constitution for the island, and in 1952, Puerto Rico became a self-governing commonwealth of the United States.

In early 2006, a financial crisis caused the government and schools to shut down for two weeks. Later that year, the government enacted the island's first sales tax to generate revenue.

Puerto Rico's most valuable farm products. The island has increased its production of livestock to help feed its growing city population.

About half of Puerto Rico's workers have service jobs. The government employs more people than any other economic activity in a variety of services, including education, medical care, and defense.

History

The economy and government of Puerto Rico are unusual in the West Indies because of the island's history. Christopher Columbus landed in Puerto Rico in 1493.

QATAR

The small Arab country of Qatar (*KUHT uhr* or *KAH tahr* or *GAH tahr*) sits on a peninsula that juts from eastern Arabia into the Persian Gulf. Most of the peninsula is a hot, stony desert, where summer temperatures rise above 120° F (49° C) and annual rainfall is rarely more than 4 inches (10 centimeters). Barren salt flats cover the southern portion of the peninsula.

For thousands of years, the people of Qatar made their living by tending camel herds, fishing, or diving for pearls in the gulf. In 1939, however, oil was discovered in western Qatar. World War II (1939-1945) delayed oil production for about 10 years, but since the 1950's, oil profits have made Qatar a rapidly developing nation.

The oil industry has also created many new jobs, and, as a result, thousands of people have moved to Qatar from other countries. In the second half of the 1900's, the population of Qatar grew 20 times larger. Today, Qatar's population is about 895,000. Arabs make up most of the population.

Most of Qatar's people live in or near Doha, the capital city, in modern houses and apartments. The majority work in Doha or other cities or in oil fields. Some Qataris wear Western clothing, but most prefer traditional Arab robes. While Arabic is the official language, many business executives and government officials speak English when they are dealing with people from other countries.

The export of petroleum and petroleum products provides the largest part of the nation's income, and Qatar ranks among the world's richest nations in terms of average income per person.

In the mid-1970's, Qatar's government took over the petroleum industry from a foreign-owned company. The government has used its oil profits to develop the nation. The government provides free health care, free housing for needy people, and free education, including special schools where adults learn to read and write. The number of schools in Qatar rose from just 1 in 1952 to about 160 in 1990.

The government has also encouraged manufacturing, fishing, and farming so that Qatar will not be entirely dependent on oil income in the future. In addition to petroleum wells and refineries, the government owns and operates a fishing fleet, flour mills, and plants that produce cement, fertilizers,

FACTS

Official name:	Dawlat Qatar (State of Qatar)
Capital:	Doha
Terrain:	Mostly flat and barren desert covered with loose sand and gravel
Area:	4,437 mi² (11,493 km²)
Climate:	Desert; hot, dry; humid and sultry in summer
Main rivers:	N/A
Highest elevation:	Tuwayyir al Hamir, 338 ft (103 m)
Lowest elevation:	Persian Gulf, sea level
Form of government:	Traditional monarchy
Head of state:	Emir
Head of government:	Prime minister
Administrative areas:	10 baladiyat (municipalities)
Legislature:	Constitution that became effective in 2005 calls for parliament with 45 members
Court system:	Courts of First Instance, Court of Appeal, Court of Cassation
Armed forces:	11,800 troops
National holidays:	Independence Day - September 3 (1971) National Day - December 18
Estimated 2010 population:	895,000
Population density:	202 persons per mi² (78 per km²)
Population distribution:	98% urban, 2% rural
Life expectancy in years:	Male, 74; female, 77
Doctors per 1,000 people:	2.6
Birth rate per 1,000:	16
Death rate per 1,000:	2
Infant mortality:	10 deaths per 1,000 live births
Age structure:	0-14: 23%; 15-64: 76%; 65 and over: 1%
Internet users per 100 people:	40
Internet code:	.qa
Languages spoken:	Arabic (official), English (widely spoken)
Religions:	Muslim 77.5%, Christian 8.5%, other 14% Note: nearly all Qatari citizens are Muslim
Currency:	Qatari riyal
Gross domestic product (GDP) in 2008:	$102.30 billion U.S.
Real annual growth rate (2008):	11.2%
GDP per capita (2008):	$121,643 U.S.
Goods exported:	Chemicals, crude oil, fertilizer, liquified natural gas, steel
Goods imported:	Iron or steel articles, machinery, transportation equipment
Trading partners:	Japan, Singapore, South Korea, United Arab Emirates, United States

Qatar lies on a desert peninsula in the Persian Gulf. In 1981, Qatar and other eastern Arabian states formed the Gulf Cooperation Council to work together on defense and economic projects.

Hawks trained to hunt small game are sold in the market at Doha. Falconry, once the "sport of kings," has been practiced in the Middle East for more than 3,000 years.

Doha is the capital and largest city of Qatar. The city's skyline features some of the world's most imaginative skyscraper architecture.

petrochemicals, plastics, and steel. Fertilizers are now an important export in addition to petroleum.

The government has dug wells to make farming possible in otherwise uncultivated areas. In addition, the government has distributed free seeds and insecticides. Qatar now produces enough vegetables for its people, though it still imports much of its meat and other food. Seawater is distilled to provide fresh drinking water.

The government of Qatar is an *emirate,* ruled by a leader called an *emir* (prince). The emir is a member of the al-Thani family, whose chiefs, or *sheiks,* became leaders of tribes in Qatar during the 1800s. The al-Thani family has ruled Qatar ever since. In 1916, Qatar became a British protectorate, but it gained complete independence in 1971. In 1972, Khalifa bin Hamad al-Thani became emir after peacefully overthrowing his cousin, Emir Ahmad bin Ali al-Thani. Khalifa bin Hamad al-Thani was in turn peacefully overthrown by his son in 1995.

In 2003, Qataris voted in favor of a new constitution, which became effective in 2005. The constitution establishes a 45-member parliament, in which 30 members are to be elected by the people and 15 members are to be appointed by the emir. The constitution allows both men and women to vote and hold office.

RÉUNION

About 400 miles (640 kilometers) east of Madagascar, in the waters of the Indian Ocean, lies the island of Réunion.

Like Mauritius, its neighbor to the east, Réunion is a volcanic island. It was formed when *magma*, or melted rock, rose from beneath the ocean floor, broke through the surface, and piled up enough to rise above the level of the water. Réunion now covers a total area of 968 square miles (2,507 square kilometers).

The interior of the island is made up of volcanic mountains with tropical rain forests scattered throughout. Réunion's highest point, Piton des Neiges, towers 10,069 feet (3,069 meters) above the forests. An active volcano, Piton de la Fournaise, stands in the southeast part of the island.

Réunion generally has a tropical climate, but temperatures are cooler at higher elevations. Rainfall also varies, from as much as 140 inches (350 centimeters) per year on the east coast, to only half that amount on the northern coast.

Portuguese sailors discovered the island in the early 1500's. It was uninhabited at the time and remained so for about 100 years. Then in 1642, the French took possession of the island and named it Bourbon. The French first used the island as a colony for prisoners. In 1665, the French East India Company, a trading firm, set up an outpost there, and the colony prospered in the coffee trade. Later, colonists used the island as a stopping-off place on the way to Mauritius. Réunion received its present name in 1848.

Since 1946, Réunion has been an overseas department of France. It is one of the few remaining possessions of France's colonial empire. The people of the island have a voice in their government, however. They elect 36 members to a governing council.

A vanilla worker sorts the valuable pods. Réunion produces much of the world's supply of vanilla, which is used as a flavoring.

Workers on a banana plantation carry their produce in a traditional way—atop their heads—along a lane shaded by tall palm trees. Many kinds of crops thrive on the island of Réunion.

Twisting roads wind between fields of sugar cane on the green hillsides of eastern Réunion. Sugar is the island's main product, accounting for most of its earnings from exports.

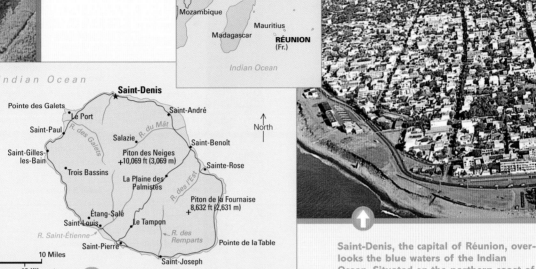

Indian Ocean

Saint-Denis

Pointe des Galets
Le Port
Saint-André
Saint-Paul
Salazie
R. du Mât
Saint-Benoît
Saint-Gilles-les-Bain
Piton des Neiges +10,069 ft (3,069 m)
Trois Bassins
Sainte-Rose
La Plaine des Palmistes
R. des l'Est
Étang-Salé
Piton de la Fournaise 8,632 ft (2,631 m)
Saint-Louis
Le Tampon
R. Saint-Étienne
R. des Remparts
Saint-Pierre
Pointe de la Table
Saint-Joseph

North

0   10 Miles
0   10 Kilometers

R. des Galets

Réunion is an island in the Indian Ocean. Of volcanic origin, it covers 968 square miles (2,507 square kilometers).

Tanzania
AFRICA
Seychelles
Mozambique
Mauritius
Madagascar
RÉUNION (Fr.)
Indian Ocean

Saint-Denis, the capital of Réunion, overlooks the blue waters of the Indian Ocean. Situated on the northern coast of the island, the town is home to about 132,000 people. Saint-Denis was founded by the French and is still quite French in character. It has many attractive buildings in the colonial style of the 1700's and 1800's, as well as spacious squares and wide avenues.

Streams of lava from one of Réunion's volcanoes provide a spectacular sight for tourists, who watch from the safety of dried lava beds. Tourism has become increasingly important to the island's economy.

The people of Réunion are largely French and African, with some Indians and Chinese. The total population of the island is about 830,000. About 132,000 people live in the capital city of Saint-Denis. Smaller towns dot the island, mainly along the shore, and a coastal road rings the entire island. A university operates in Saint-Denis.

Tourism has become increasingly important to the economy of Réunion. The French government has invested heavily in the tourist industry on the island and has improved its network of roads, including the coastal road. The molten lava flows, pleasant tropical climate, and beautiful beaches are all tourist attractions.

Many of the people of Réunion, however, are farmers. The island's most important agricultural product is sugar. Farmers also grow tea, tobacco, vanilla, and flowers.

Réunion produces much of the world's supply of vanilla from orchids that grow on the island. These climbing orchids grow on a vine that attaches itself to trees. The plant produces pods, or beans, that are gathered when they are yellow-green in color, then dried, or cured. This process shrinks the pod and changes its color to a rich chocolate-brown. This curing process also gives the vanilla bean its familiar flavor and aroma.

Réunion's tropical flowers are also the source of the perfumes produced on the island. Fragrant plants have tiny *sacs* (baglike parts) that make and store the substances that give them their pleasant scent. Workers on Réunion extract these substances—called *essential oils*—from flower petals as well as other parts of plants to make perfumes and colognes.

ROMANIA

Romania—also spelled *Rumania* or *Roumania*—is a country in eastern Europe. The country was a monarchy from 1881 until the end of World War II (1939-1945). In 1947, it became a Soviet satellite with a Communist government controlled by the U.S.S.R. During the 1950's, Romanian leaders began to oppose Soviet intervention, and they insisted that Romania be allowed to make its own foreign policy.

Nicolae Ceaușescu, who had served as general secretary and leader of the Romanian Communist Party since 1965, became head of state in 1968. He continued Romania's efforts to free itself from Soviet control. During this time, Romania entered a period of rapid industrial expansion. Industry—including manufacturing, mining, and construction—passed agriculture as the leading producer of income in Romania.

During an economic crisis in the early 1980's, the Romanian government declared that it would pay back all its foreign debts. To meet its payments, Romania dramatically increased its exports while decreasing its imports, causing shortages of food and consumer goods for the Romanian people. Meanwhile, Ceaușescu's luxurious lifestyle stood in dramatic contrast to the poverty of the people. In addition, a report issued by the human rights organization Amnesty International in 1987 condemned the government's restrictions on the lives of the Romanian people.

In 1988, Ceaușescu began a widely criticized program of "rural urbanization." The plan called for the destruction of about 8,000 Romanian villages and the resettlement of the villagers into multistory housing complexes. The economic hardships, the human rights abuses, and the rural urbanization plan helped bring down Ceaușescu's Communist regime in 1989. Revolutionaries seized control of the radio and television stations and formed the National Salvation Front (NSF).

During the revolution, Ceaușescu and his wife were captured near Tîrgoviște as they were trying to escape. They were convicted of various charges—including mass murder, corruption, and the destruction of Romania's economy—and later executed.

FACTS

Official name:	Republica Romania (Republic of Romania)
Capital:	Bucharest
Terrain:	The plain and plateau of Transylvania is separated from the plain of Moldavia on the east by the Carpathian Mountains and separated from the Walachian plain on the south by the Southern Carpathian Mountains
Area:	92,043 mi² (238,391 km²)
Climate:	Temperate; cold, cloudy winters with frequent snow and fog; sunny summers with frequent showers and thunderstorms
Main rivers:	Arges, Danube, Ialomiţa, Jiu, Olt, Prut, Siret
Highest elevation:	Mount Moldoveanu, 8,343 ft (2,543 m)
Lowest elevation:	Black Sea, sea level
Form of government:	Republic
Head of state:	President
Head of government:	Prime minister
Administrative areas:	41 judete (counties), 1 municipiu (municipality)
Legislature:	Parlament (Parliament) consisting of the Senat (Senate) with 137 members serving four-year terms and the Adunarea Deputatilor (Chamber of Deputies) with 334 members serving four-year terms
Court system:	Supreme Court of Justice
Armed forces:	73,200 troops
National holiday:	Unification Day - December 1 (1918)
Estimated 2010 population:	21,150,000
Population density:	230 persons per mi² (89 per km²)
Population distribution:	55% urban, 45% rural
Life expectancy in years:	Male, 68; female, 76
Doctors per 1,000 people:	1.6
Birth rate per 1,000:	10
Death rate per 1,000:	12
Infant mortality:	13 deaths per 1,000 live births
Age structure:	0-14: 15%; 15-64: 70%; 65 and over: 15%
Internet users per 100 people:	25
Internet code:	.ro
Languages spoken:	Romanian (official), Hungarian, German
Religions:	Eastern Orthodox 86.8%, Protestant 7.5%, Roman Catholic 4.7%, other 1%
Currency:	New Romanian leu
Gross domestic product (GDP) in 2008:	$200.07 billion U.S.
Real annual growth rate (2008):	7.6%
GDP per capita (2008):	$9,298 U.S.
Goods exported:	Chemicals, clothing, iron and steel, machinery, petroleum products, transportation equipment
Goods imported:	Chemicals, oil and petroleum products, electrical equipment, food, machinery, motor vehicles
Trading partners:	France, Germany, Hungary, Italy, Russia, Turkey, Ukraine, United States

An agricultural country before the Communists gained control, Romania has always been one of Europe's least developed nations. As a result of the industrialization program of the 1960's, industry has surpassed agriculture in economic importance.

The Palace of the Parliament in Bucharest contains both chambers of the Romanian Parliament. The giant palace has been called the world's largest civilian administrative building.

Multiparty presidential and legislative elections held in May 1990 gave the NSF an overwhelming victory. Ion Iliescu, leader of the NSF, became president.

In mid-1990, pro- and antigovernment demonstrators clashed in Bucharest, injuring hundreds. In late 1991, following strikes and riots by miners, the NSF government resigned. It was replaced by a coalition dominated by the NSF.

In general elections in 1992, the Democratic National Salvation Front (DNSF), a leftist group that had split from the NSF earlier in the year, won the majority of the seats in the legislature. Throughout all the turmoil, Iliescu remained president.

Romania's economy worsened, its national debt expanded, and it attracted little foreign investment. In 1994, wages fell and many workers were jobless. In 1996, Romania elected its first non-Communist government since the fall of the Ceauşescus in 1989. Iliescu was reelected again in 2000.

In 2004, Romania joined NATO. Later that same year, opposition leader Traian Basescu was elected as president of Romania. In 2007, Romania joined the European Union (EU).

In April 2007, Romania's parliament accused Basescu of abuse of power and suspended him from office. One month later, voters rejected his suspension and returned him to power. Basescu was reelected to a second term in 2009.

ENVIRONMENT

Lying deep in the heart of eastern Europe, Romania is bordered to the north by Ukraine, to the east by Moldova, to the west by Hungary and Serbia, and to the south by Bulgaria. The Black Sea lies along the country's southeast coast for 130 miles (209 kilometers), and the Danube River flows through Romania for about 900 miles (1,400 kilometers).

Romania has a landscape of striking contrasts. The mountain region of the northern and central part of the country features breathtaking scenery, hiking trails, and many ski and vacation resorts. The region's numerous tiny lakes add a special touch of beauty. Picturesque villages lie scattered across the vast fertile farmlands surrounding the mountains, while sandy beaches line the sunny coast of the Black Sea.

Romania's many important natural resources include fertile cropland, rich pastureland, forests, and valuable mineral deposits in the mountains and plateaus. Grains, especially corn and wheat, are Romania's leading crops. Grapes and other fruits, potatoes, and sugar beets are grown in the fertile soil of the plains and plateaus. Farmers raise more sheep than any other kind of livestock. They also raise cattle, horses, pigs, and poultry.

The country's six land regions are Transylvania, Bukovina, Moldavia, Walachia, Banat, and Dobruja.

Transylvania

Transylvania, the largest and most diverse of Romania's land regions, includes most of the country's mountains, the Transylvanian Plateau, and the northwestern plain. Beautiful towns, churches, villages, and farms are set among its highlands and broad valleys.

The great arc of the Carpathian Mountain System encircles the Transylvanian Plateau. The Moldavian, or Eastern, Carpathians extend from the northern border to the center of the country, while the Transylvanian Alps, or Southern Carpathians, stretch westward from the Moldavian range. The Bihor Mountains and other ranges make up the Western Carpathians.

On the Moldavian plain, women in traditional dress stack hay to dry in the sun. Farmers grow grains and fruits in Moldavia's fertile soil.

Houses and farm buildings are clustered in a valley in the Transylvanian Alps. Mount Moldoveanu, Romania's highest mountain, is located in the Transylvanian Alps. It rises 8,343 feet (2,543 meters).

"Dracula's Castle"—the stronghold of the cruel prince from Walachia who inspired the legend—stands high atop a mountain in the Transylvanian Alps in central Romania. The character of Dracula is based on Vlad Tepes, a prince who committed hundreds of savage murders in the 1400's and executed many of his enemies by driving a sharpened pole through their bodies. The novel *Dracula* (1897), by the English author Bram Stoker, made the legend internationally famous.

The Carpathians are neither very high nor steep, and the many passes cut through them enable tourists to explore the mountains and to observe foxes, lynx, badgers, wolves, and even bears in their natural habitat. Hikers enjoy following tracks over the peaks and cliffs and through gorges, dense forests, and mountain meadows.

Today, Transylvania may be best known as the home of Count Dracula—the wicked nobleman in the famous vampire story. Northeast of Transylvania in the Moldavian Carpathian Mountains lies the thickly forested region of Bukovina.

Plains and plateaus

In the regions of Moldavia, Walachia, and Banat, the land descends from mountains near Transylvania to hills and then to plains that provide Romania's best farmland. Moldavia forms part of the Russian steppes, but the original steppe vegetation has disappeared with crop cultivation. The city of Bucharest, Romania's chief industrial center, is located in Walachia.

Dobruja, a small plain between the northern course of the Danube River and the Black Sea, is home to an abundance of wildlife, including hundreds of species of birds. Sturgeon, the source of caviar, are also found in the waters of the Danube where the river empties into the Black Sea.

Herds of cattle are watered in a stream in Walachia. Livestock accounts for about two-fifths of the value of Romania's agricultural products, and crops account for about three-fifths. Walachia has more people than any other Romanian region.

Fishing crews sail downriver at the Mouths of the Danube. East of the city of Galaţi, the Danube forms a three-pronged delta consisting of rivers, islands, lakes, ponds, and streams. Wildlife in the region is now threatened by pollution.

PEOPLE

The Romanians are the only people in eastern Europe who trace their ancestry and language back to the Romans. Today, a vast majority of the people in Romania are descended from the Dacians and Romans and from tribes such as the Goths, Huns, and Slavs.

Early migrations

The Dacians lived in what is now Romania before the Romans arrived. By the 300's B.C., they were farming the land, mining gold and iron ore from the mountains, and trading with neighboring peoples.

In A.D. 106, the Romans, led by Emperor Trajan, conquered the Dacians and made the region a province of the Roman Empire. Roman soldiers occupied Dacia, and Roman colonists settled there. The Dacians intermarried with the Romans and adopted their customs and the Latin language. The region, which had previously been called *Dacia*, became known as *Romania*.

Barbarians arriving from the east and north during the 200's forced the Romans to abandon the province. From the late 200's to the 1100's, a series of invaders swept through Romania, including Bulgars, Goths, Huns, Magyars, Slavs, and Tatars. Like the Romans, these groups—especially the Slavs—also intermarried with the Romanians.

The Romanian language, which developed from Latin, is not at all like other languages in eastern Europe. Romanian most closely resembles French, Italian, Portuguese, and Spanish—languages that also developed from Latin.

Minority groups

Romania's largest minority group is the Hungarians, who make up about 7 percent of the population and live mostly in Transylvania. The region in which the ethnic Hungarians now live consists of territory taken from Hungary after World War I (1914-1918).

Romania's government has faced criticism for not adequately recognizing the ethnic and cultural

Religious worship was suppressed during the period of Communist rule in Romania. However, many people continued to attend church services and celebrate religious festivals. Most Romanians belong to the Romanian Orthodox Church, an Eastern Orthodox church.

Rural Romanians often celebrate weddings, christenings, and religious holidays by wearing traditional costumes. These festive celebrations are the most important part of social life in rural Romania.

People of German origin are a significant minority group in Romania. Hungarians form the largest minority group. Schools in Romania provide lessons in Hungarian and German as well as in Romanian.

The Beer Cart is one of Bucharest's best-known and most popular restaurants. The Beer Cart is decorated to resemble a traditional German beer garden. It opened in 1879.

A fisherman propels his flat-bottomed boat through the dense reeds in Romania's delta region. An increasing number of Romania's people are moving from rural areas to the cities to take jobs in industry.

rights of the Hungarians in Romania. In March 1990, several people died in clashes between ethnic Hungarians and Romanian nationalists in Tîrgu Mures.

Other minority groups include Germans, Jews, Turks, Ukrainians, and Roma (sometimes called Gypsies). In 1919, German settlers in Transylvania and part of Banat agreed to join the new Romanian state, which promised them many rights and freedoms. After World War II, however, many ethnic Germans fled or were expelled from Romania. Their departure, along with Romania's poor economic conditions, left many German settlements in a state of decay. Big country estates fell into disrepair, soil was left untilled, and churches were locked up and boarded.

Romanians have one of the lowest living standards in Europe. Although almost all Romanian workers earn enough to pay for basic needs, few can afford luxury items. Rapid population growth has caused a housing shortage in the cities, and most city people live in crowded apartments.

RUSSIA

From the gently rolling European Plain to the vast Siberian wilderness, and from the soaring Ural Mountains to the depths of Lake Baikal, the huge country of Russia spreads across the two continents of Europe and Asia. It is a nation whose long and tumultuous history has shaped world affairs as few others have.

The historical roots of the Russian people date from the early migrations of the Slavs, who had established settlements in what is now the European part of Russia and who became known as the Eastern Slavs. About 988, Grand Prince Vladimir I, the ruler of the Eastern Slavs, converted to Christianity, and most people under his rule became Christians.

As Russia expanded its territory and absorbed other nationality groups, its rulers forced the conquered peoples to adopt Orthodox Christianity and the Russian language. These rulers were known as *czars* (emperors), and they created a centralized state that had complete power over every aspect of Russian life.

Under the czars, the country was largely cut off from the industrial progress made in Western Europe in the 1800's, and most of its people remained poor, uneducated peasants. However, despite their difficult life, the peasants loved their country, which they called "Mother Russia."

In 1917, Czar, Nicholas II was driven from power in what is now known as the February Revolution. In 1922, the Communist leaders, who had taken over the Russian government, formed the Union of Soviet Socialist Republics (U.S.S.R.), or Soviet Union. The Soviet Union initially consisted of Russia plus three other republics. By 1940, 12 other republics had joined—or were forcibly annexed to—the Soviet Union, making it the largest and most powerful Communist country in the world.

After the establishment of the Soviet Union, the Russians continued to dominate other nationality groups throughout the union republics. Soldiers, bureaucrats, and teachers carried out "Russification" programs. In the 1930's, as workers moved to non-Russian republics during Joseph Stalin's industrialization program, Russian influence spread even wider.

In 1991, Communist rule in the Soviet Union collapsed, and the country broke apart. Russia and most of the other republics formed a new, loose federation called the Commonwealth of Independent States. After the breakup of the Soviet Union, Russia's new national government worked to move the country from a state-controlled economy to one based on private enterprise. The government also began to establish new political and legal systems in Russia.

109

RUSSIA TODAY

Russia was the most important republic in the Soviet Union. When the Soviet Union broke apart in 1991, Russia entered a difficult period that promised a new beginning for the Russian people.

When the Communists first took over the country in 1917, Russia's people were poor and uneducated. The Communists sought to expand the economy under a series of plans emphasizing industrialization. As a result, the Soviet Union became the world's largest centrally planned economy. The Soviet government owned most of the nation's banks, factories, land, mines, and transportation systems. It also planned and controlled the production, distribution, and pricing of almost all goods.

However, the focus on heavy manufacturing caused frequent shortages of consumer goods. As a result, improvements in living conditions came slowly, and there were often shortages of food and other basic household necessities. In addition, the Communists restricted people's basic freedoms. Those who opposed government policies were often expelled from their homes, sent to prison labor camps or mental hospitals, or executed.

In 1985, Mikhail S. Gorbachev became the leader of the Soviet Union. He instituted many reforms, including a policy of openness that became known as *glasnost*. Meanwhile, popular movements in Russia and the other republics of the Soviet Union threatened the unity of the nation. Gorbachev called for a new treaty that would grant the republics a large amount of independence.

On Aug. 19, 1991—the day before the treaty was to be signed—conservative Communist officials tried to overthrow Gorbachev's government. Boris N. Yeltsin, president of the Russian republic, led an uprising against the coup, which quickly collapsed. On Dec. 17, 1991, Yeltsin and Gorbachev agreed to dissolve the Soviet Union and replace it with the Commonwealth of Independent States, an association of former Soviet republics.

In the early 1990's, Yeltsin's government supported major economic reforms, including a rapid shift to private enterprise. However, his policies brought inflation and a recession, and he faced opposition from parliament. Yeltsin accused the parliament of resisting reform and disbanded it in 1993. Russia's voters elected a new parliament and approved a new constitution.

FACTS

Official name:	Rossiyskaya Federatsiya (Russian Federation)
Capital:	Moscow
Terrain:	Broad plain with low hills west of Urals; vast coniferous forest and tundra in Siberia; uplands and mountains along southern border regions
Area:	6,601,669 mi² (17,098,242 km²)
Climate:	Ranges from steppes in the south through humid continental in much of European Russia; subarctic in Siberia to tundra in the polar north; winters vary from cool along Black Sea coast to frigid in Siberia; summers vary from warm in the steppes to cool along Arctic coast
Main rivers:	Lena, Ob, Volga
Highest elevation:	Mount Elbrus, 18,510 ft (5,642 m)
Lowest elevation:	Coast of Caspian Sea, 92 ft (28 m) below sea level
Form of government:	Republic
Head of state:	President
Head of government:	Prime minister
Administrative areas:	46 oblasts, 21 republics, 4 autonomous okrugs, 9 krays, 2 federal cities, 1 autonomous oblast
Legislature:	Federalnoye Sobraniye (Federal Assembly) consisting of the Sovet Federatsii (Federation Council) with 168 members serving four-year terms and the Gosudarstvennaya Duma (State Duma) with 450 members serving four-year terms
Court system:	Constitutional Court
Armed forces:	1,027,000 troops
National holiday:	Russia Day - June 12 (1990)
Estimated 2010 population:	140,542,000
Population density:	21 persons per mi² (8 per km²)
Population distribution:	73% urban, 27% rural
Life expectancy in years:	Male, 60; female, 73
Doctors per 1,000 people:	4.3
Birth rate per 1,000:	11
Death rate per 1,000:	15
Infant mortality:	11 deaths per 1,000 live births
Age structure:	0-14: 15%; 15-64: 71%; 65 and over: 14%
Internet users per 100 people:	21
Internet code:	.ru
Languages spoken:	Russian, many minority languages
Religions:	Russian Orthodox, Muslim, Protestant, Buddhist, Jewish, Roman Catholic, Hindu
Currency:	Russian ruble
Gross domestic product (GDP) in 2008:	$1.680 trillion U.S.
Real annual growth rate (2008):	6.0%
GDP per capita (2008):	$11,888 U.S.
Goods exported:	Chemicals, machinery, metals, natural gas, petroleum and petroleum products, wood products
Goods imported:	Chemicals, food and beverages, machinery, motor vehicles
Trading partners:	Belarus, China, Germany, Italy, Japan, Netherlands, Ukraine

North Pole
Arctic Ocean
Chukchi Sea
Bering Strait
UNITED STATES (Alaska)
Bering Sea
Wrangel I.
Longa Strait
Chukchi Peninsula
Gulf of Anadyr
UNITED KINGDOM
Norwegian Sea
North Sea
Franz Josef Land
East Siberian Sea
Anadyr
Cape Navarin
NORWAY
DENMARK
SWEDEN
Barents Sea
Severnaya Zemlya
New Siberian Islands
Anadyr Range
Bilibino
Koryak Mountains
Cape Olyutorskiy
Murmansk
Kola Peninsula
FINLAND
Kandalaksha
Novaya Zemlya
Vilkitski Strait
Cape Chelyuskin
Laptev Sea
Laptev Strait
Kolyma Lowland
Eastern Siberia
Baltic Sea
Gulf of Finland
RUSSIA
Kaliningrad
LATVIA
ESTONIA
Segezha
White Sea
Kara Sea
Taymyr Peninsula
Khatanga Gulf
Yanskiy Gulf
Karagin I.
Commander Is.
POLAND
LITHUANIA
Pskov
Ladoga
Severodvinsk
St. Petersburg
Arkhangelsk
Kara Strait
Yenisey Gulf
Taymyr Lake
Yamal Peninsula
North Siberian Lowland
Mt. Pobeda 10,325 ft. (3,147 m)
Shelikhov Gulf
Ust-Kamchatsk
Velikiy Novgorod
BELARUS
Velikiye Luki
Onega
Konosha
Vyshniy Volochek
Pechora Basin
Gydan Peninsula
Cherskiy Range
Ust-Nera
Susuman
Kamchatka Pen.
Klyuchevskaya 15,584 ft. (4,750 m)
Smolensk
Tver
Rybinsk Res.
Pechora
Vorkuta
Norilsk
Putorana Plateau
Verkhoyansk Range
Magadan
Moscow
Yaroslavl
Ukhta
Mt. Narodnaya 6,217 ft. (1,895 m)
Tazovskaya Bay
Igarka
Petropavlovsk-Kamchatskiy
Bryansk
Tula
Ivanovo
Gorki Res.
Syktyvkar
Northern Uvals
Ural Mountains
Vilyuy Range
Sea of Okhotsk
Cape Lopatka
UKRAINE
Kursk
Orel
Ryazan
Nizhniy-Novgorod
Kirov
Central Siberian Plateau
Yakutsk
Ust-Maya
Lipetsk
Tambov
Solikamsk
Sergino
West Siberian Plain
Voronezh
Kuybyshev Res.
Perm
Serov
Surgut
Lower Tunguska
Kempendyay
Dzhugdzhur Range
Russia claims and occupies the Kuril Islands. But Japan also claims the southernmost Kurils.
Udskaya Bay
Okha
Kuril Islands
Kazan
Izhevsk
Kama
Nizhnevartovsk
Yenisey
Mirnyy
Aldan
Sea of Azov
Penza
Ulyanovsk
Samara
Yekaterinburg
Lensk
Aldan Mts.
Rostov-on-Don
Saratov
Tolyatti
Tavda
Tobolsk
Severo Yeniseyskiy
Aleksandrovsk-Sakhalinskiy
Tsimlyansk Res.
Ufa
Tyumen
Stanovoy Range
Black Sea
Krasnodar
Volgograd Res.
Ust-Ishim
Belyy Yar
Bodaybo
Neryungri
Tatar Strait
Sakhalin Island
Volga-Don Canal
Volgograd
Magnitogorsk
Chelyabinsk
Kolpashevo
Lesosibirsk
Angara
BAM LINE
Dzhagdy Mts.
Zeya
Komsomolsk
Sovetskaya Gavan
Mt. Elbrus 18,510 ft. (5,642 m)
Elista
Stavropol
Orenburg
Tomsk
Ust-Ilimsk
Ust-Kut
Yuzhno-Sakhalinsk
GEORGIA
Caspian Lowland
Astrakhan
Orsk
Omsk
Ob Res.
Novosibirsk
Krasnoyarsk
Krasnoyarsk Res.
Bratsk
Bratsk Res.
TRANS-SIBERIAN RAILROAD
Khabarovsk
Sikhote-Alin Range
Grozny
CHECHNYA
Caucasus Mts.
TURKEY
Makhachkala
ARMENIA
Barnaul
Novokuznetsk
Kulunda Steppe
Sayan Mts.
Irkutsk
Lake Baikal
Chita
Blagoveshchensk
KAZAKHSTAN
Kuznetsk Basin
Abaza
Ulan-Ude
Lake Khanka
AZERBAIJAN
Kyzyl
Erzin
Mt. Munku-Sardyk 11,453 ft. (3,491)
CHINA
Mt. Belukha 14,783 ft. (4,506 m)
Caspian Sea
Vladivostok
Nakhodka
UZBEKISTAN
MONGOLIA
Sea of Japan (East Sea)
JAPAN
TURKMENISTAN
IRAN
500 Miles
500 Kilometers
North
NORTH KOREA

RUSSIA

Russia is the world's largest country in area. It is almost twice the size of the second largest country, Canada. Russia extends from the Arctic Ocean south to the Black Sea and from the Baltic Sea east to the Pacific Ocean.

In 1994, Russian troops invaded the *autonomous* (self-governing) republic of Chechnya to put down a separatist movement. Russia invaded Chechnya again in 1999 to defeat a rebellion in which Islamic militants attempted to unite Chechnya and the neighboring republic of Dagestan. Many nations protested Russia's handling of the conflict.

Despite political turmoil and ill health, Yeltsin was reelected president in 1996. He resigned in 1999 and appointed Vladimir Putin, the prime minister, as acting president. In elections in 2000, Russians voted overwhelmingly for Putin. He was reelected in 2004. Under Putin's leadership, Russia developed friendlier relations with Europe and the United States. In 2002, Russia entered into a special partnership with the North Atlantic Treaty Organization (NATO), a military alliance that had been formed to oppose the Soviet Union.

Political problems persisted, however. In 2004, Chechen terrorists seized an elementary school in Beslan, a town in southwest Russia. More than 350 adults and children were killed as Russian security forces stormed the school to free the hostages.

In 2008, Russia and Georgia clashed over control of South Ossetia, a region in north-central Georgia, and Russian troops entered Georgia. Russia recognized South Ossetia and Abkhazia, another region of Georgia, as independent nations. However, the United States and other Western nations considered the regions to be part of Georgia. European Union observers monitored Russia's withdrawal from Georgia.

In 2008, Dmitry Medvedev, Putin's handpicked successor, became president in elections that many observers considered unfair. As expected, Medvedev chose Putin to serve as prime minister. That same year, Medvedev pushed through constitutional reforms lengthening the presidential term of office from four years to six. The move was widely viewed as part of a plan for Putin to return to power.

Putin was reelected president in 2012 despite widespread peaceful protests. The protesters questioned the legitimacy of Putin's possible 12-year rule and called for an end to political corruption.

EARLY HISTORY

The history of Russia began many centuries ago, when the broad steppes of the southern region formed a natural corridor for migrating peoples. By the A.D. 800's, Slavic groups had built many towns in what is now the European part of Russia and Ukraine. The first state they founded was called Kiev Rus, and the city of Kiev became its capital.

According to the *Primary Chronicle,* the earliest written Russian history, a group of Vikings called the *Varangian Russes* captured Kiev in 882. By the 900's, the other Russian *principalities* (regions ruled by a prince) recognized Kiev's importance as a cultural and commercial center.

During the 1200's, *Tatar* (Mongol) armies swept across Russia from the east, destroying one town after another, and in 1240, when the Mongols destroyed Kiev, Russia became part of the Mongol Empire. Under Mongol control, Russia was cut off from the influence and new ideas of the Renaissance—an important cultural movement that dramatically changed Western Europe and later its overseas colonies.

The first czar

During the early 1300's, the Mongols grew weak, and the small northeast principality of Moscow became rich and powerful. By 1480, Moscow's leader, Ivan III (Ivan the Great), had ended Mongol control of Russia. In 1547, Ivan IV (Ivan the Terrible) became the first ruler to be crowned czar.

A crafty and cruel leader, Ivan the Terrible laid the foundation for the growth of Russia. Under his rule, Russian forces crossed the Ural Mountains to conquer western Siberia. They also took control of the region along the Volga River. Ivan passed a series of laws that bound the peasants to the land as *serfs*—making them part of the landowner's property.

After Ivan's death in 1584, Russia was torn by civil war, foreign invasion, and political confusion until 1613. That year, an assembly of nobles and citizens elected Michael Romanov czar. His descendants ruled Russia for three centuries.

Under the leadership of Czar Peter I (Peter the Great), who ruled from 1682 to 1725, Russia came to be a major European power. Peter expanded Russian territory to the Baltic Sea and introduced many Western ways to the nation. His succes-

Ivan the Terrible
(1530-1584)

Peter the Great
(1672-1725)

Catherine the Great
(1729-1796)

TIMELINE

A.D. 800's	Slavic groups settle in what is now the European part of Russia and Ukraine. Eastern Slavs establish the state of Kiev Rus.
c. 988	Grand Prince Vladimir I converts to Christianity.
1237	Batu, grandson of Genghis Khan, leads Mongol forces into Russia.
1240	Mongols destroy Kiev; Russia becomes part of the Mongol Empire.
c. 1318	Prince Yuri of Moscow is appointed the Russian grand prince.
1380	Grand Prince Dmitri defeats a Mongol force in the Battle of Kulikovo.
Late 1400's	Moscow becomes the most powerful Russian city.
1480	Ivan III (Ivan the Great) breaks Mongol control over Russia.
1547	Ivan IV (Ivan the Terrible) becomes the first ruler to be crowned czar.
1554	Russian forces conquer western Siberia.
1556	Ivan IV defeats Astrakhan.
1604-1613	Russia is torn by civil war, invasion, and political confusion during the Time of Troubles.
1613	Michael Romanov becomes czar.
1600's	Russia extends control to Ukraine and Siberia.
1682-1725	Peter I (Peter the Great) reigns as czar.
1703	Peter the Great founds St. Petersburg.
1709	Russia defeats Sweden in the Great Northern War.
1762-1796	Empress Catherine II (Catherine the Great) rules Russia.
1773-1774	Russian troops crush a peasant revolt.
Late 1700's	Russia gains parts of Poland, the Crimea, and other Turkish lands.
1812	Napoleon I leads French army to Moscow, but is forced to retreat.
1825	Government troops crush the Decembrist uprising.
1853-1856	Russia fights the Ottoman Empire in the Crimean War and is defeated.
1861	Alexander II frees the serfs.
1904-1905	Japan defeats Russia in the Russo-Japanese War.
1905	A revolution in January forces Nicholas II to establish a parliament; a general strike in October paralyzes the country.
1914-1917	Russia fights Germany and Austria-Hungary in World War I.

Alaska was a territory of Russia from the 1700's to 1867.

Boundary of Moscow 1462
Expansion 1462-1533
Expansion 1533-1584
Expansion 1584-1689
Expansion 1689-1914
Boundary of present-day Russia

Between 1462 and 1914, Russia gained vast new territories through conquests and annexations. The boundary of present-day Russia appears as a solid red line.

The Trinity Monastery of St. Sergius at Sergiyev Posad, northeast of Moscow, served as a treasury of Russian art and literature for centuries.

In 1905, thousands of unarmed workers marched to the czar's Winter Palace in St. Petersburg to demand political and social reforms. Government troops fired into the crowd, killing or wounding hundreds of people.

sors, including Empress Catherine II (Catherine the Great), continued to promote Western culture and ideas in Russia. Lavish parties took place at the czar's palace, the arts were encouraged, and many new schools were established. But the great majority of the Russian people continued to live in extreme poverty, and the landowners kept tight control over the serfs.

Harsh rule continued in Russia until Alexander II, who reigned from 1855 to 1881, introduced major reforms. In 1861, Alexander freed the serfs, distributed land among the peasants, and established forms of self-government in the towns and villages. However, many young Russians felt that Alexander's new policies did not go far enough. After a revolutionary tried to kill Alexander in 1866, the czar began to weaken his reforms. He was killed by a terrorist bomb in 1881, and his son, Alexander III, began a program of harsh rule.

The coming of revolution

In 1894, Nicholas II became Russia's last czar. When a series of bad harvests caused widespread starvation, discontent grew among the rising middle class and workers in the cities. Various political organizations soon emerged. After an economic depression began in 1899, student protests, peasant revolts, and worker strikes increased. However, aside from making a few constitutional reforms—such as forming a Duma (parliament)—the czar and his officials refused to give up power.

During World War I (1914-1918), the Russian economy could not support both the war effort and the people at home. As a result, severe shortages of food, fuel, and housing occurred throughout the nation. In March 1917, the Russian people revolted. When soldiers who were called in to halt the uprising sided with strikers and demonstrators, the days of the Russian czars were ended forever.

MODERN HISTORY

By March 9, 1917, about 200,000 Russian workers were demonstrating in Petrograd (now St. Petersburg), then the capital of Russia. In the midst of the riots and strikes, which became known as the February Revolution, the Duma established a *provisional* (temporary) democratic government, and a new Soviet of Workers' and Soldiers' Deputies was formed in Petrograd in March. Nicholas II, who had lost all political support, gave up the throne on March 15.

The October Revolution

Neither the provisional government nor the Soviet of Workers' and Soldiers' Deputies was powerful enough to rule on its own. In late 1917, the *Bolshevik* (majority) wing of the Russian Social Democratic Labor Party, under the leadership of Vladimir I. Lenin and Leon Trotsky, seized power and formed a new Russian government. They spread Bolshevik rule through the local *soviets* (councils).

The new government soon faced a civil war against antirevolutionary forces aided by troops from France, the United Kingdom, Japan, the United States, and other countries opposed to the Bolsheviks' Communist policies. The civil war between the *Reds* (the Bolsheviks, renamed the Russian Communist Party) and the *Whites* (the anti-Communist forces) lasted from 1918 to 1920, when the Reds defeated the poorly organized Whites.

By 1921, seven years of war, revolution, civil war, and invasion had exhausted Russia and severely disrupted its economy. In an effort to deal with the growing discontent and new uprisings, the government introduced the New Economic Policy (NEP), which permitted small industries and the retail trade to operate on their own.

In 1922, the new Russian Republic became one of the four founding republics of the Union of Soviet Socialist Republics (U.S.S.R.). Russia was the U.S.S.R.'s largest and most powerful republic.

The rise of Stalin

By the mid-1920's, as a result of Lenin's New Economic Policy, all the nation's factories and other means of production were operating again. Lenin died in 1924, and Joseph Stalin, who had become general secretary of the

Boris Yeltsin leads the opposition to a coup against Soviet leader Mikhail S. Gorbachev from the top of a tank in Moscow in 1991. The coup failed, and the Soviet Union rapidly dissolved. Yeltsin led Russia through the difficult transition period that followed.

A study in the Winter Palace at Petrograd (now St. Petersburg), which served as the headquarters of the provisional government in 1917, stands in shambles after it was stormed by armed workers and Bolsheviks during the October Revolution.

Communist Party's Central Committee, began his rise to supreme power. By 1929, Stalin had become dictator of the Soviet Union.

Under Stalin's First Five-Year Plan, farmers were forced to give most of their products to the government at low prices. But the farmers, resisting Stalin's orders, destroyed much of their livestock and crops. This action caused widespread starvation, and Stalin sent several million peasant families to prison labor camps in Siberia and Soviet Central Asia.

During the 1930's, Stalin began a program of terror called the Great Purge, in which his secret police arrested anyone and everyone suspected of being a threat to his power. Up to 20 million people may have been killed during this period.

During World War II (1939-1945), following the German invasion of the U.S.S.R., the Soviet Union became a partner of the Allies. After World War II, Stalin gradually cut off almost all contact between the U.S.S.R. and Western nations.

TIMELINE

1917	Nicholas II is overthrown in the February Revolution; the Bolsheviks, led by V. I. Lenin, take over the government in the October Revolution.
1918-1920	Communist forces (Reds) defeat anti-Communist forces (Whites) in civil war.
1922	The Union of Soviet Socialist Republics (U.S.S.R.) is established; Joseph Stalin becomes general secretary of the Communist Party.
1924	V. I. Lenin dies.
1928	The First Five-Year Plan begins.
1929	Stalin defeats his political rivals and becomes dictator of the U.S.S.R.
Mid-1930's	Millions of Soviet citizens are imprisoned or executed during the Great Purge.
1941	German forces invade the Soviet Union during World War II.
1942-1943	Soviet troops prevent the Nazis from capturing Stalingrad.
1945	The Soviets' capture of Berlin leads to a German surrender.
Late 1940's	The Soviet Union gains control in eastern Europe by setting up satellite states with Communist governments. The Iron Curtain falls and the Cold War develops.
1953	Stalin dies; Nikita Khrushchev becomes head of the Communist Party.
1957	Soviet scientists launch the satellite Sputnik I.
1961	Yuri Gagarin becomes the first person to orbit Earth.
1962	The Cuban missile crisis occurs.
1964	Khrushchev is replaced by Leonid I. Brezhnev as head of the Communist Party.
1982	Brezhnev dies. Yuri V. Andropov becomes Communist Party head.
1984	Andropov dies. Konstantin U. Chernenko replaces Andropov.
1985	Chernenko dies. Mikhail S. Gorbachev becomes head of the Communist Party and announces a policy of *glasnost* (openness) and *perestroika* (restructuring).
1990	Gorbachev is awarded the Nobel Peace Prize.
1991	Communist rule ends, and the Soviet Union is dissolved. Russia and other Soviet republics become independent nations. Chechnya begins drive for independence from Russia.
Early 1990's	Russia, under President Boris Yeltsin, undergoes major economic reforms, including a rapid shift to private enterprise.
2000	Vladimir Putin becomes president.
2008	Dmitry Medvedev becomes president with Putin as prime minister; Medvedev signs a law extending the president's term from four to six years in a move widely seen as paving the way for Putin's return to the presidency; Russia and Georgia clash over control of South Ossetia.
2012	Putin is reelected president amid widespread, peaceful protests and allegations of voting irregularities.

At the Yalta Conference in February 1945, key Allied leaders met to address the major problems in a postwar Europe. Left to right: Winston Churchill, Franklin D. Roosevelt, and Joseph Stalin.

Post-Stalin

Soviet relations with the West did not improve until after Stalin's death in 1953. His successor, Nikita Khrushchev, announced a policy of "peaceful coexistence" and eased some restrictions on communication, trade, and travel. Khrushchev was replaced by Leonid I. Brezhnev, who also pursued a policy of friendlier relations with the West. He was succeeded by Yuri V. Andropov, who was later replaced by Konstantin U. Chernenko.

Mikhail S. Gorbachev came to power in 1985. Under him, the Soviet Union changed rapidly, then dissolved. In December 1991, the world recognized Russia as an independent nation led by Boris N. Yeltsin.

The Union of Soviet Socialist Republics (U.S.S.R) consisted of 15 republics, including Russia. It covered more than half of Europe and nearly two-fifths of Asia. The U.S.S.R. dissolved in 1991, and the republics became independent nations.

Vladimir I. Lenin (1870-1924)

Joseph Stalin (1879-1953)

Vladimir Putin (1952-)

PEOPLE AND WAY OF LIFE

The people of Russia are distributed unevenly throughout the country. The vast majority live in the western—or European—part of Russia. The more rugged and remote areas to the east are sparsely inhabited.

About 80 percent of Russia's people are of Russian ancestry. Members of more than 100 other nationality groups also live in Russia. The largest groups include Tatars (or Tartars), Ukrainians, Chuvash, Bashkirs, Belarusians, Mordvins, Chechens, Germans, Udmurts, Mari, Kazakhs, Avars, Armenians, and Jews, who are considered a nationality group in Russia.

The government of the Soviet Union controlled many aspects of life for the Russian people. It exerted great influence over religion, education, and the arts. The independence of Russia following the breakup of the Soviet Union brought greater freedom and triggered many other changes in the lives of the people.

Under Communism

When the Communists took over the Russian government in 1917, they hoped to create a classless society—one with neither rich people nor poor people. They set out to take control of the economy, the educational system, and the cultural life of the people. However, the Communists failed to achieve that goal. Under the Soviet system, privileged groups enjoyed special rights, while the vast majority of Soviet citizens had a significantly lower standard of living.

Soviet citizens suffered great restrictions on their personal freedom. People who criticized the country's political system were severely punished, or even killed. During the 1930's—the years of Stalin's Great Purge—the government's secret police arrested millions of Soviet citizens suspected of anti-Communist views or activities. These people were either shot or sent to labor camps in Siberia.

A St. Petersburg couple examines a bulletin board advertising apartments available. Most apartments are very small and cramped.

Ballet dancers perform to enthusiastic audiences at the Kirov Theater in St. Petersburg and the Bolshoi Theater in Moscow (right). A Russian family (below) enjoys St. Petersburg's Hermitage, a former czarist palace that contains one of the world's great collections of art.

A huge display of wild mushrooms is spread out in a stall at Moscow's private market. Russian people often sell food collected in the wild or grown on small, private plots.

The Communists destroyed many churches and persecuted religious leaders after they rose to power in 1917. However, many people continued to worship in private. In 1990, the government ended all religious restrictions.

A new openness

In the late 1980's, Mikhail Gorbachev's policy of glasnost led to greater freedom for the Soviet people. The government relaxed its control of newspapers, radio, and television. Greater religious freedom was permitted, and a wave of cultural activity swept across the country. After the Communist Party fell from power in 1991, party officials and other privileged groups lost their special rights.

Although glasnost and the suspension of the Communist Party created a new spirit of openness in the U.S.S.R., many serious problems have remained in the independent state of Russia. Most of the people live in urban areas, and cities are very crowded. Millions of families live in small, plain, crowded apartments, and housing shortages frequently force many families to share apartments.

Shortages of food, services, and manufactured goods have been common features of city life in Russia. The shift toward capitalism that began in the 1990's has not yet cured the shortages. Even when goods become available, they are often too expensive for many people to afford. Russian cities also face such urban problems as crime and environmental pollution.

In rural areas, single-family housing is common. In the most remote areas of Russia, some homes lack gas, plumbing, running water, and electric power. In addition, the quality of education, health care, and cultural life is lower than in the cities. Rural life is changing, however. Rural stores, for example, have a wider selection of goods available than they once offered.

The Soviet government controlled education and considered it a major vehicle of social advancement. As a result, almost all Russians can read and write. Today, public education in Russia remains free for all citizens. New private schools have also opened. Russian educators are changing the school curriculum to better prepare students for the new economy.

The Russian Orthodox Church is the largest religious denomination in the country. In addition to Russian Orthodoxy, religions that have full freedom in Russia include Buddhism, Islam, Judaism, and certain Christian denominations. These religions enjoy full freedom because they were recognized by the state prior to the fall of the Soviet Union.

Workers leave an automobile factory in the city of Nizhniy Novgorod. Heavy industry remains the most important part of the Russian economy.

THE VOLGA RIVER

Flowing for about 2,300 miles (3,700 kilometers) through the heart of Russia, the Volga River ranks as the longest river in European Russia. From the earliest days of Slavic settlement, the Volga has served as an important trade and communications route. The Lena River in Siberia, 2,734 miles (4,400 kilometers) long, is Russia's longest river.

Long ago, Viking sailors navigated the Volga, trading furs and slaves for silver and silks from Asia. Since that time, many key events in Russian history have taken place along the river's shores. The beauty and significance of *Matushka Volga* (Mother Volga) have long been celebrated in the country's music and literature.

A major waterway

The Volga River rises in the Valai Hills, between Moscow and St. Petersburg, at an elevation of only about 748 feet (228 meters) above sea level. It first flows eastward, passing to the north of Moscow, and then southwest after passing the city of Kazan. South of Volgograd, the river turns southeast before discharging its waters into the Caspian Sea through a 100-mile (160-kilometer) delta. At the Caspian Sea, the Volga is 92 feet (28 meters) below sea level.

With its many tributaries, the Volga forms a major river system that drains an area of about 525,000 square miles (1,360,000 square kilometers). Farmers grow wheat and other crops in this fertile river valley, which is also rich in petroleum, natural gas, salt, and potash. The delta region and the Caspian Sea provide one of the world's great fishing grounds.

The river begins to freeze at the end of November, and in some areas, the ice does not clear until April. When it is not frozen, the Volga, including many of its tributaries, is navigable for almost its entire length. Canal networks, such as the Volga-Don Canal in the south, link the river with the Baltic, White, and Black seas, and its tributaries carry timber to the Volga from as far away as the Urals. The Volga carries a large amount of the river traffic in Russia, including steamships that transport passengers as well as freight.

The Volga begins in the forested countryside and flows through Tver, the first major city on its banks. Later, the river swells from its many tributaries, including the Oka, the Kama, the Vetluga, and the Sura rivers.

Situated along the banks of the Volga, the city of Volgograd, once known as Stalingrad, was the site of an important Soviet victory during World War II. Today, the city is a major industrial center and an important stop for traffic on the river.

Pollution problems

In its journey to the Caspian Sea, the Volga passes through some of the most densely populated areas of Russia. Unfortunately, the intensive development of manufacturing, mining, petroleum, and modern agriculture in the river basin has created serious pollution problems.

Industrial and agricultural activity uses huge amounts of river water, much of which is discharged back into the river. In addition, numerous dams slow the river's flow, reducing its natural ability to purify itself.

Dams also prevent the sturgeon from migrating to its normal spawning ground. The eggs, or *roe,* of these large freshwater fish are known as *caviar,* one of the region's most famous and valuable food products. To solve the migrating problem, authorities have set up artificial hatcheries, where they produce sturgeon to stock the river.

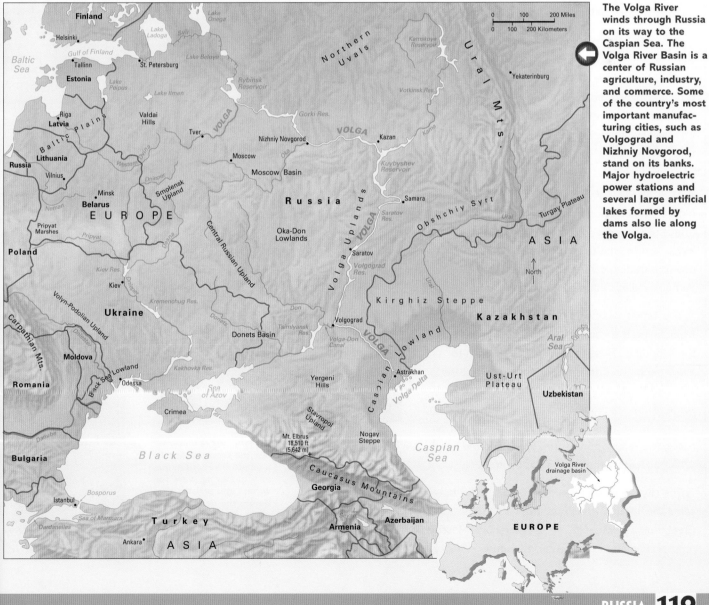

The Volga River winds through Russia on its way to the Caspian Sea. The Volga River Basin is a center of Russian agriculture, industry, and commerce. Some of the country's most important manufacturing cities, such as Volgograd and Nizhniy Novgorod, stand on its banks. Major hydroelectric power stations and several large artificial lakes formed by dams also lie along the Volga.

MOSCOW

Ranking among the largest cities in the world, Moscow is the capital of Russia. The city lies in the western part of Russia, about 400 miles (640 kilometers) southeast of St. Petersburg. More than 10 million people live in this great city.

From the air, Moscow looks like a huge wheel, with boulevards extending from the center like spokes. Circular roads cross the boulevards, forming inner and outer rims. Moscow is a city of extraordinary cultural spectacles, fascinating historical sights, and breathtaking architecture. From the famous Bolshoi Ballet, whose dancers are considered by many to be the most skilled and graceful in the world, to the huge GUM department store, Moscow dazzles visitors with its many sights and sounds.

An old Russian proverb states, "Above Russia there is only Moscow; above Moscow, only the Kremlin; and above the Kremlin, only God." Many Russian citizens dream of living in this historic city, but the severe shortage of housing means that no one may live there without official permission from the government.

Early history

Moscow was founded in 1147 by Yuri Dolgoruki, a prince of the region. Lying along the banks of the Moscow River, the town grew wealthy and prosperous as a trading center. Then, during the 1200's, Mongols called Tatars conquered the area.

The Tatars, who were primarily interested in maintaining their power and collecting taxes from the conquered Russian principalities, began to allow the grand prince of Moscow to collect the taxes for them. Ivan I, the prince of Moscow, kept some of the tax money and used it to buy land and expand his territory. As Moscow grew stronger and richer, the Tatars' power grew weaker. In the late 1400's, Moscow threw off Tatar control.

The city grew rapidly in the 1600's, and it became the home of the czars. Even after Peter the Great built a new capital at St. Petersburg, Moscow remained an important center of culture, industry,

Shoppers throng GUM, the largest department store in Russia. This shopping complex, which stands on the site of a historic market, displays a wide range of goods, from clothing to caviar.

Graceful street lanterns line the crowded Arbat, a Moscow street famous as a gathering place for writers and artists. Before the days of glasnost, the Arbat was a refuge for intellectuals who disagreed with government policies.

and trade. In 1812, French armies led by Napoleon I reached Moscow, but a mysterious fire—believed by some to have been set by the Russians themselves—destroyed most of the city.

In 1918, the Bolsheviks moved the capital back to Moscow, and the city was once again the nation's political center. In 1991, Moscow became the center of protests that helped end the coup attempted by conservative Communist officials.

A center for industry and culture

Moscow is the most important industrial city in Russia. Its factories produce a wide variety of products, including automobiles, buses and trucks, chemicals, electrical machinery, measuring instruments, steel, and textiles.

The city's many important educational and cultural institutions include the State Historical Museum, which attracts many students of Russian history. The Russian State Library is the largest library in Russia, and one of the largest libraries in the world. Two of the best-known universities in the city are M. V. Lomonosov Moscow State University and the Peoples' Friendship University of Russia.

Perhaps the most impressive of Moscow's buildings is the Kremlin, the fortified enclosure within the city. With its many gilded domes, its tapered gate towers, and the contrast between its forbidding walls and the beauty of its interior, it offers one of the most breathtaking sights in the world.

The walls of the Kremlin tower over Red Square in the heart of Moscow. The Kremlin is a massive fortress housing the Russian government as well as museums, magnificent palaces, and cathedrals. Its triangular enclosure extends almost 1-1/2 miles (2.4 kilometers) around. The Kremlin's present walls have stood since the late 1400's.

A Moscow street vendor in a fur cap sells tulips. Private enterprise began to flourish in Moscow in the late 1980's as a result of President Gorbachev's policy of perestroika.

Central Moscow includes many major historical and cultural sites. At the heart of the city stands the Kremlin, a one time fortress that is now the center of the Russian government. A large plaza called Red Square lies just outside the Kremlin walls. The Moscow River weaves through the city.

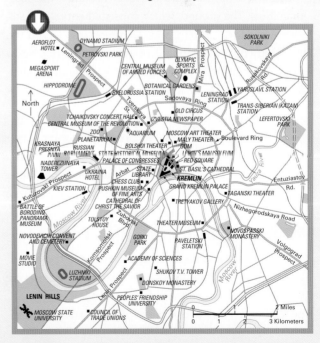

ST. PETERSBURG

St. Petersburg, the second largest city in Russia, lies in the northwestern part of the country, at the eastern end of the Gulf of Finland. A major port and one of the world's leading cultural and industrial centers, St. Petersburg is a magnificent city of luxurious palaces, handsome public buildings, fine museums and theaters, and wide public squares resembling those of the great Western European cities.

From 1924 until 1991, the city was known as *Leningrad*—after the Communist leader Vladimir I. Lenin. In 1991, however, as Communist influence declined in the country, Leningrad's citizens voted in a nonbinding referendum to restore the city's original name of St. Petersburg. In September 1991, the Soviet government officially approved the change.

Russia's "window to the West"

The city was founded in 1703 by Peter the Great. Peter was the first Russian czar to visit Western Europe, and he was impressed by what he saw there. He returned from his travels determined to bring Western culture and technology to Russia. He promptly ordered the building of a "window to the West"—a city that would be a showcase for his efforts to westernize Russian life.

The site he selected—a marshy lowland where the Neva River empties into the Gulf of Finland—was less than ideal. The ground, which consisted entirely of silt, was subject to flooding, and the region also lacked the necessary building supplies of stone and timber. However, Peter saw that the area would provide good river links to the interior if it had a canal system.

Forced laborers, including prisoners of war, were brought in from across the empire to drain the marshes and dig the canals. French and Italian architects were hired to design and erect buildings and churches. Peter then forbade the building of stone houses elsewhere in Russia, saying that all the nation's stonemasons were needed in the new city.

The results were astonishing. Where once there was a vast, deserted swampland, there stood a dazzling city whose beauty rivaled that of Paris, London, and Vienna. The city boasted some of the finest examples of

Baroque and Neoclassical architecture in the world. Wide boulevards, known as *prospekts,* had been built along the canals, and splendid palaces outside the city—such as Pavlovsk, Peterhof, and Pushkin—became the summer residences for the Russian imperial family.

Peter moved the nation's capital from Moscow to St. Petersburg in 1712, and the new capital soon became the intellectual and social center of the Russian Empire. Many of the greatest Russian writers of the 1800's lived and worked in St. Petersburg. Even today, its nearly 5 million residents are proud of their city's cosmopolitan atmosphere and Western outlook.

War and revolution

Over the years, St. Petersburg has served as a backdrop for many of the great events in Russian history. When Russia went to war against Germany at the outbreak of World War I (1914-1918), the city's name was changed from St. Petersburg to Petrograd, to avoid the German ending of *burg.* After the February Revolution of 1917, the beautiful Winter Palace in the center of the city became the headquarters of the provisional government.

The Winter Palace, completed in 1762, was the main residence of Russia's monarchs. St. Petersburg served as the capital of Russia from 1712 to 1918. The Winter Palace is now part of the State Hermitage Museum. It faces the Neva River.

The golden dome of the Cathedral of St. Isaac of Dalmatia—along with the golden spire of the Admiralty Building and the green-and-white Hermitage Museum—grace the St. Petersburg skyline. In the distance stand the city's factories.

A single shot from the cruiser *Aurora* opened the October Revolution of 1917, when the Bolsheviks seized power in what was then Petrograd. Moored in the Neva River, the ship now serves as a museum. St. Petersburg's other museums include the Hermitage Museum world famous for its magnificent works of art.

The Peterhof Palace, on the outskirts of St. Petersburg, was once the summer residence of the czars. The palace was restored to its former splendor after being almost completely destroyed during World War II.

Only a few months later, the Bolsheviks seized Petrograd during the October Revolution, and they moved the capital back to Moscow in 1918. When Lenin died in 1924, Petrograd was renamed Leningrad. In 1934, Sergey Kirov, a prominent Soviet leader, was assassinated in the city—an event that triggered Stalin's Great Purge.

During World War II (1939-1945), the Germans laid siege to the city from September 1941 to January 1944. About a million people died during the siege, mostly from starvation. Much of Peter the Great's dazzling city lay in ruins. After the war, however, many beautiful and historic buildings were lovingly restored and rebuilt, and St. Petersburg is once again the spectacular city it was in imperial times.

SIBERIA

A vast, thinly populated region of Russia, Siberia covers some 4,938,000 square miles (12,789,000 square kilometers) in northern Asia. The region is bounded by the Ural Mountains to the west, the Pacific Ocean to the east, the Arctic Ocean to the north, and Kazakhstan, Mongolia, China, and North Korea to the south. The historical area of Siberia is divided into three economic regions—West Siberia, East Siberia, and the Far East.

The region's abundant natural resources make Siberia extremely valuable to the Russian economy. Oil fields along the Ob River produce a large percentage of the nation's oil, and much of its coal comes from deposits in the Kuznetsk Basin. Siberian forests provide large supplies of timber. In addition, the region has extensive natural gas reserves near the Arctic Circle and around Yakutsk.

Environment

The *tundra*—a cold, dry region where no trees can grow—lies along Siberia's Arctic coast. The only vegetation consists of mosses, lichens, grasses, low shrubs, and grasslike plants called *sedges*. Few people live in the tundra, and arctic foxes, lemmings, and reindeer roam freely.

South of the tundra, the evergreen forests of the *taiga* stretch from the Urals all the way to the Pacific. The taiga's wildlife includes ermines, lynxes, red foxes, and sables. The *steppes* (grasslands), in the extreme southwest, contain the region's richest farmland.

Siberia's temperatures are among the coldest on Earth, and the region's harsh climate has always been a barrier to settlement. In northeastern Siberia, average temperatures range from below –50° F (–46° C) in January to about 60° F (16° C) in July. Siberian summers are brief, and the growing season is short—from 90 to 150 days. In most of eastern Siberia, the ground remains frozen for many months of the year. Along the Pacific coast, the winters are wet and stormy.

Siberia, which stretches from the Ural mountains on the west to the Pacific Ocean on the east, makes up about 20 percent of the area of Russia, cent of the Russian people live there.

The Indigirka River, in northeast Siberia, winds through a landscape of evergreen forests and bogs typical of the taiga. Most of the trees in this region are needleleaf evergreens. Hydroelectric dams on Siberia's rivers provide a small portion of Russia's electric power.

Land regions

Siberia's three main land regions are the West Siberian Plain, the Central Siberian Plateau, and the East Siberian Highlands. Extending from the Ural Mountains to the Yenisey River, the West Siberian Plain ranks as the world's largest flat region. Its landscape is broken only by a few *moraine* hills formed from glacial deposits. Because the land is so flat, rain water drainage is poor, and the area is covered with swamps and marshes.

The Central Siberian Plateau lies between the Yenisey and Lena rivers. Because the plateau is cut by deep river valleys, it often appears mountainous—particularly in the north and west, where the rivers have carved steep canyons through thick layers of volcanic rock.

The East Siberian Highlands consist of a series of mountain ranges between the Lena River and the Pacific coast. In the north, the Verkhoyansk Range and the Cherskiy Range form a huge crescent that rises 10,000 feet (3,350 meters) in some areas.

In the northeast, the Koryak Mountains make up the spine of the Kamchatka Peninsula. This peninsula has about 25 active volcanoes, including Klyuchevskaya, the highest point in Siberia at 15,584 feet (4,750 meters). Along the Chinese border to the south, the Yablonovyy and Stanovoy ranges form a rugged highland region that reaches to the Pacific Ocean.

Ice and snow cover most of Siberia for about six months a year, and the temperature sometimes drops below −90° F (−68° C). However, temperatures and snowfall vary widely from west to east.

LAKE BAIKAL

Lake Baikal, the world's deepest lake, lies on the southern edge of the Central Siberian Plateau. Although the lake is 395 miles (636 kilometers) long and 49 miles (79 kilometers) wide, its surface is frozen from January to May. The large volume of water in Lake Baikal affects the weather in the area surrounding it. The area nearest the lake is several degrees warmer in winter, and cooler in summer, than places farther away from the moderating influence of the lake.

PEOPLE OF SIBERIA

Although Siberia covers about 75 percent of Russia, less than 20 percent of the Russian people live there. The region's harsh weather, poor soil, and rugged terrain have long discouraged settlers. Today, Siberia has as a population of about 37 million. It has a population density of only 7-1/2 people per square mile (3 people per square kilometer).

Early settlement

A group of Asian nomads called Tatars, led by the Mongol conqueror Genghis Khan, invaded the southern steppes of Siberia in the 1200's. They drove the region's original inhabitants into the northern forests. In the late 1500's, the Tatars in southwestern Siberia were conquered by a band of Russian Cossacks.

Gradually, the influence of the Russians spread eastward into the forests. Russian fur traders reached the Pacific coast about 1630, and the settlement of Okhotsk was founded in 1649. By 1700, Russia controlled nearly all of Siberia.

Many of the early Russian settlers were drawn by the profits to be made in trapping the numerous fur-bearing animals of the Siberian forests—particularly sables. Many others migrated to Siberia in hopes of escaping religious persecution and the miserable conditions in the overcrowded farmlands of European Russia.

Soon, the Russian czars began using the Siberian forests as a place of exile for political opponents. After Russia became part of the Soviet Union, Joseph Stalin followed the same policy in the 1930's. He banished millions of Soviet citizens to labor camps in the Siberian wilderness.

Many political exiles were forced to work as laborers in Siberian mines, factories, and construction projects. Siberia's population grew during World War II (1939-1945), as the Soviet government moved hundreds of factories and thousands of workers east of the Urals to protect them from invading German armies. When Stalin died in 1953, the use of forced labor in Siberia ended.

Buryat women on a collective farm near Ulan-Ude represent one of Siberia's major native groups. The Buryats are descendants of the Mongols and live in their own autonomous republic east of Lake Baikal.

Siberians today

Most present-day Siberians are Russian, and many are descendants of settlers and fur traders who arrived in the 1600's. However, many people whose ancestors were the original inhabitants of Siberia still maintain their old territories and way of life.

The Buryats, a group of North Mongolians, and the Yakuts, who live in eastern Siberia, have their own autonomous republics. About 1,500 Yuit live on the northeastern tip of Siberia and make their living by herding reindeer, hunting walruses and other animals, and selling carvings and other handicrafts. The Evenki hunt and herd reindeer in the central and eastern Siberian tundra and taiga.

Snow blankets a village near Lake Baikal, but life for the villagers goes on as usual. About 30 percent of the Siberian people live in rural areas. Many have simple log houses in the expansive wilderness.

Melons offered for sale in a market in Irkutsk is imported from warmer regions. Siberian crops consist mostly of barley, oats, and wheat grown on the southern steppes during the short summer growing season.

About 70 percent of Siberia's people live in cities, where they are crowded into small apartments. Many people in rural areas live in simple, but more spacious, log houses. Novosibirsk, the largest city in Siberia, has a population of about 1-1/2 million. Other large Siberian cities include Omsk, Krasnoyarsk, Vladivostok, Irkutsk, Barnaul, Khabarovsk, Novokuznetsk, and Kemerovo.

Living standards in Siberia are lower than elsewhere in Russia, and many Siberians complain of boredom and a lack of cultural activities. In addition, working conditions in the bitter cold of Siberia are often extremely difficult.

In an effort to attract workers to Siberia, the Russian government has offered high salaries and long vacations, but many people stay only a few years. However, others accept the hardships more readily, knowing that social and professional advancement come more quickly in Siberia than elsewhere in Russia.

High wages lure workers like these Mongolians to Siberia, but living conditions can be both harsh and tedious. Most of the profits from Siberian products flow west, leaving little behind for developing the area or establishing cultural activities for the people.

THE TRANS-SIBERIAN RAILROAD

Every morning, a train known as *Rossiya No. 2* leaves Moscow's Yaroslavl Station for Vladivostok, a port on Russia's Pacific coast. In its seven-day journey across the great Siberian wilderness, the train travels more than 5,000 miles (8,000 kilometers) and crosses seven time zones. Known as the Trans-Siberian Express, the train travels along the same route as the original Trans-Siberian Railroad line—the longest railroad in the world.

Building the railroad

The Trans-Siberian Railroad was the first railroad built across the vast region of Siberia. Before the railroad was constructed, the only route across Siberia was a trail of mud and dust in the summer and snow in the winter. The route was traveled by few people except convicts, bandits, and political exiles.

After approving the construction of the great railroad, Czar Alexander III added "Most August Founder of the Great Siberian Railway" to his many titles. Begun in 1891 and finished in 1916, the Trans-Siberian Railroad was built in several sections. The section in eastern Siberia, between Vladivostok and Khabarovsk, was completed about 1897. By 1904, a continuous railroad stretched from Vladivostok across China and Siberia to the Ural Mountains.

Railroad workers take a break at a small station between Irkutsk and Khabarovsk.

A samovar (coal-heated stove) provides a constant supply of hot water for tea, an increasing popular drink on board the Trans-Siberian Express.

Whistling past a small community in the taiga, the Trans-Siberian Express gives passengers an ever-changing view of life on Russia's vast frontier. The train's final stop is the Pacific port of Vladivostok.

The Trans-Siberian Railroad is the longest railroad in the world, extending more than 5,600 miles (9,000 kilometers). It stretches across the width of Russia, from Moscow in the west to Vladivostok at the far southeastern tip of the country.

Moscow
Nizhniy Novgorod
Ural Mts.
Yekaterinburg
Siberia
Novosibirsk
Krasnoyarsk
Baikal-Amur Main Line
Severobaikalsk
L. Baikal
Irkutsk
Ulan-Ude
Tynda
Sovetskaya Gavan
Khabarovsk
Vladivostok

— Trans-Siberian Railroad
— Trans-Siberian Railroad serviced connection rail line

0   1,000 Miles
0   1,000 Kilometers

However, the Russians wanted a route that did not cross China, so in 1916 they completed a line north of China from Khabarovsk to Kuenga—the last link in a continuous railroad on Russian soil. Since the 1920's, the Trans-Siberian has been joined to other railroads in neighboring republics that once composed the Soviet Union. The Trans-Siberian route is now part of the rail network that links this vast territory spanning Europe and Asia.

Prisoners and laborers

In the early years of building the railroad, migrant farmers and peasant settlers provided most of the labor. As construction progressed and more workers were needed, the Railway Committee began using thousands of convicts, social misfits, and political prisoners who had been banished to Siberia. These prisoners earned a reduced sentence in return for their labors.

The work was very difficult, and conditions were both harsh and hazardous. The engineers lacked heavy machinery, and workers had only wooden shovels to break through the frozen ground. Winters were bitterly cold, and during the summer, flies and mosquitoes from the swamps were a constant torment. The railroad workers also had to battle disease, bandits, and even the Amur tiger.

All aboard

The 1914 edition of *Baedeker's Guide to Russia* advised travelers on the Trans-Siberian Railroad "to carry a revolver in Manchuria and on trips away from the railway." Though conditions have improved over time, the accommodations on the Trans-Siberian Express are less than luxurious by Western standards.

Boarding the train, passengers are greeted by a *provodnik*—an attendant whose many duties include making sure passengers are not left behind during stopovers; tending the *samovar* (a coal stove used to heat water for tea); and scraping ice off the steps of the train. Each car in the first-class section has a toilet and washbasin. However, washing up can be an awkward experience due to the small size of the area and the constant movement of the train. And all first- and second-class passengers must make up their own berths.

Powerful, Czech-built electric locomotives haul the Trans-Siberian Express between Moscow and Irkutsk.

RWANDA

Rwanda is a small country in east-central Africa, just south of the equator. With about 10½ million people, it is one of the most crowded countries on the continent. Because it has little industry and more people than its land area can support, Rwanda is also one of the world's poorest countries. The nation has long been troubled by ethnic conflict.

People and economy

A large majority of Rwandans belong to the Hutu (also called Bahutu) ethnic group. The Tutsi (also called Batutsi or Watusi) form a minority of the population. Both groups speak Kinyarwanda, a Bantu language. The Twa, a Pygmy group, make up less than 1 percent of the population.

Most Rwandans are farmers, but many can grow only enough food to feed their families. Their food crops include bananas, beans, cassava, potatoes, sorghum, and sweet potatoes. Some farmers also raise cattle and goats. Coffee is the country's chief export. Most manufacturing in Rwanda is for domestic consumption. The country has no railroads.

History and government

Hutu farmers and Pygmy hunters were the first known inhabitants of what is now Rwanda. About 600 years ago, the Tutsi invaded from the north and took control.

Germany conquered the area that is now Rwanda and Burundi—the country to the south—in 1897. Belgium later took control of the region, called Ruanda-Urundi. The death of the region's king, Mwami Mutara III, in 1959 led to violence. The Hutu rebelled against the Tutsi, and thousands died. The Hutu then gained control of the government.

In 1961, the people of Ruanda-Urundi voted to make their country a republic. Ruanda-Urundi became two independent nations—Rwanda and Burundi—on July 1, 1962. Gregoire Kayibanda was elected as Rwanda's first president.

Kayibanda was overthrown by military officers led by Major General Juvenal Habyarimana, a Hutu, in 1973. Habyarimana then became president, established a single political party, and imposed a new constitution.

In 1990, a rebel group called the Rwandan Patriotic Front (RPF), consisting mostly of Tutsi, began launching attacks against the government. In April 1994, Habyarimana died after his airplane was shot down. Hutu extremists in

FACTS

Official name:	Republika y'u Rwanda (Republic of Rwanda)
Capital:	Kigali
Terrain:	Mostly grassy uplands and hills; relief is mountainous with altitude declining from west to east
Area:	10,169 mi² (26,338 km²)
Climate:	Temperate; two rainy seasons (February to April, November to January); mild in mountains with frost and snow possible
Main rivers:	Rusizi, Kagera, Akanyaru, Nyabarongo, Mwogo
Highest elevation:	Karisimbi, 14,787 ft (4,507 m)
Lowest elevation:	Rusizi River, 3,117 ft (950 m)
Form of government:	Republic
Head of state:	President
Head of government:	Prime minister
Administrative areas:	4 provinces, 1 city
Legislature:	Chamber of Deputies with 80 members serving five-year terms and Senate with 26 members serving eight-year terms
Court system:	Supreme Court
Armed forces:	33,000 troops
National holiday:	Independence Day - July 1 (1962)
Estimated 2010 population:	10,534,000
Population density:	1,036 persons per mi² (400 per km²)
Population distribution:	81% rural, 19% urban
Life expectancy in years:	Male, 48; female, 50
Doctors per 1,000 people:	Less than 0.05
Birth rate per 1,000:	43
Death rate per 1,000:	16
Infant mortality:	86 deaths per 1,000 live births
Age structure:	0-14: 43%; 15-64: 55%; 65 and over: 2%
Internet users per 100 people:	3
Internet code:	.rw
Languages spoken:	Kinyarwanda (official), French (official), English (official), Swahili
Religions:	Roman Catholic 56.5%, Protestant 26%, Adventist 11.1%, Muslim 4.6%, other 1.8%
Currency:	Rwandan franc
Gross domestic product (GDP) in 2008:	$4.46 billion U.S.
Real annual growth rate (2008):	7.5%
GDP per capita (2008):	$467 U.S.
Goods exported:	Coffee, hides, niobium, tea, tin
Goods imported:	Food, machinery, motor vehicles, petroleum products, pharmaceuticals, steel
Trading partners:	Belgium, China, Kenya, Tanzania, Uganda

Rwanda's government then began a campaign of violence. Hutu militias massacred thousands of Tutsi and moderate Hutu. The RPF launched attacks in response. Many Hutu fled to the Democratic Republic of the Congo, at that time called Zaire. The RPF defeated the Hutu forces and took control of Rwanda's government. Pasteur Bizimungu, a Hutu in the RPF, was appointed president.

In November 1994, the United Nations created a special court of justice to prosecute the organizers of the *genocide* (systematic killing) of Tutsi and moderate Hutu. The RPF government sought to promote reconciliation, but ethnic tensions remained high.

From 1994 to 1996, Hutu extremists took control of many of the refugee camps in Zaire and joined Zairian government forces in attacking Tutsi. Rwanda responded by attacking the refugee camps. Rwanda also aided rebels in the overthrow of Zaire's government in 1997. The rebels renamed Zaire the Democratic Republic of the Congo. In 1998, Rwandan troops backed another group of rebels in a war against Congo's new government.

Paul Kagame, a Tutsi, became president in 2000. In 2002, Congo and Rwanda signed a peace agreement. In 2003, Rwandan voters approved a new constitution. That same year, Kagame was elected to a seven-year term as president. He was reelected in 2010. In 2007, Rwanda joined the East African Community, which aims to promote economic and political cooperation among its members.

Acres of crops surround the farm buildings on a palm oil plantation. Agriculture is the main economic activity in Rwanda. Coffee is the nation's chief export.

Rwanda is a small country in east-central Africa, just south of the equator. The country lies on a high plateau. Rwanda's landscape ranges from volcanic mountains to winding river valleys, and from beautiful lakes to grassy plains.

A colorful headdress is worn by a young woman of the Hutu, an ethnic group that was dominated by the minority Tutsi until 1959.

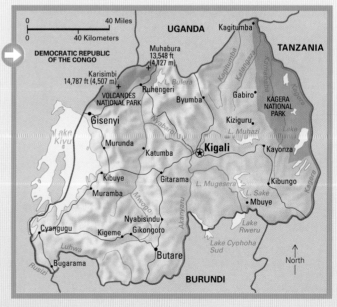

SAINT KITTS AND NEVIS

St. Kitts and Nevis is an independent country consisting of two islands—St. Kitts and Nevis. The islands are about 2 miles (3.2 kilometers) apart, separated by a waterway known as *the Narrows*. St. Kitts and Nevis lie about 190 miles (310 kilometers) east of Puerto Rico, and they have a land area of 101 square miles (261 square kilometers). They are part of the Leeward Islands group of the Lesser Antilles.

Both St. Kitts and Nevis are the tops of volcanic mountains that rise out of the Caribbean Sea, and many beaches on the islands consist of black volcanic sand. A narrow strip of fertile plains lies along the coasts.

The Carib Indians, who inhabited the islands before the arrival of European explorers and settlers, called St. Kitts *Liamuiga*, or *Fertile Land*. When Christopher Columbus sighted the island in 1493, he named it after his patron saint, Saint Christopher. British settlers later called the island *St. Kitts*. The name *Nevis* comes from *nieve*, the Spanish word for *snow*, because the Spaniards mistook the clouds on Nevis Peak for snow.

St. Kitts and Nevis consists of two islands. St. Kitts covers 65 square miles (168 square kilometers). Nevis covers 36 square miles (93 square kilometers).

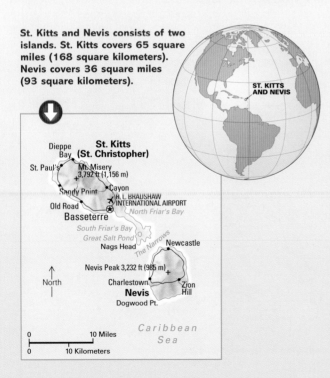

FACTS

Official name:	Federation of Saint Kitts and Nevis
Capital:	Basseterre
Terrain:	Volcanic with mountainous interiors
Area:	101 mi² (261 km²)
Climate:	Tropical tempered by constant sea breezes; little seasonal temperature variation; rainy season (May to November)
Main rivers:	N/A
Highest elevation:	Mount Misery, 3,792 ft (1,156 m)
Lowest elevation:	Caribbean Sea level
Form of government:	Constitutional monarchy
Head of state:	British monarch, represented by governor general
Head of government:	Prime minister
Administrative areas:	14 parishes
Legislature:	National Assembly with 14 members serving five-year terms
Court system:	Eastern Caribbean Supreme Court
Armed forces:	N/A
National holiday:	Independence Day - September 19 (1983)
Estimated 2010 population:	51,000
Population density:	505 persons per mi² (195 per km²)
Population distribution:	68% rural, 32% urban
Life expectancy in years:	Male, 69; female, 74
Doctors per 1,000 people:	1.1
Birth rate per 1,000:	18
Death rate per 1,000:	8
Infant mortality:	14 deaths per 1,000 live births
Age structure:	0-14: 27%; 15-64: 65%; 65 and over: 8%
Internet users per 100 people:	30
Internet code:	.kn
Language spoken:	English (official)
Religions:	Anglican, other Protestant, Roman Catholic
Currency:	East Caribbean dollar
Gross domestic product (GDP) in 2008:	$551 million U.S.
Real annual growth rate (2008):	3.5%
GDP per capita (2008):	$12,530 U.S.
Goods exported:	Electronics, food and beverages, machinery
Goods imported:	Electronics, food, iron and steel, machinery, motor vehicles, petroleum products
Trading partners:	Canada, Trinidad and Tobago, United Kingdom, United States

Lush tropical growth covers the slopes of Nevis Peak, which rises 3,232 feet (985 meters) in the center of the island of Nevis.

The light of the setting sun is reflected in the waters of the Caribbean Sea, as waves gently lap against a quiet beach in Anguilla. Anguilla is a coral island north of St. Kitts.

British colonists began settling on St. Kitts in 1623. The settlement was Britain's first possession in the West Indies. Soon, French colonists also settled on St. Kitts. The British colonized Nevis in 1628. The early settlers grew tobacco on the islands but switched to growing sugar cane after the prices for tobacco fell on the European market. The settlers established large plantations and imported African slaves to work on them. Today, almost all the people of St. Kitts and Nevis are descendants of Africans.

The British gained control of the entire island of St. Kitts in 1713. French troops captured it during the American Revolution (1775-1783), but British ownership of the island was recognized in the peace treaty signed in 1783. The mighty Brimstone Hill fortress, built by slave labor over the course of almost 100 years, made St. Kitts the "Gibraltar of the West Indies."

In 1816, Anguilla became part of the same colony as St. Kitts and Nevis. In 1967, the three islands became an associated state of the United Kingdom. Anguilla became a separate British dependency in 1980, and St. Kitts and Nevis became an independent nation in 1983. In 1998, voters on Nevis narrowly defeated a referendum that would have made the island independent from St. Kitts.

Basseterre, the capital of St. Kitts and Nevis, lies on the southwestern coast of St. Kitts. French settlers gave Basseterre its name, and its old colonial buildings give the city a distinctly French flavor.

A vendor sells fruit at a street market in Basseterre, the capital and chief urban center of St. Kitts and Nevis. The fertile land of Nevis is divided into small farms that produce fruits and vegetables.

ANGUILLA

The island of Anguilla first became a British colony in 1650. It lies about 70 miles (113 kilometers) north of St. Kitts and Nevis, with which it once formed a single British colony. The island's hot climate and erratic rainfall are not suitable for commercial agriculture. Instead, fishing and tourism are the mainstays of the economy.

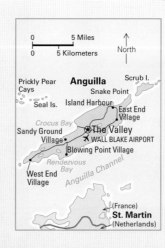

SAINT LUCIA

St. Lucia is a small island country in the Lesser Antilles, lying north of St. Vincent and south of Martinique. Some St. Lucians claim that Christopher Columbus discovered their island in 1502, during his fourth voyage to the New World. But other evidence suggests that Columbus sailed past the island and an unknown explorer discovered it later.

The Arawak Indians were the original inhabitants of St. Lucia, but they were conquered by the Carib Indians in the 1300's. During the early 1600's, the Carib resisted attempts by the French and British to colonize the island. Finally, in the mid-1600's, the French succeeded in setting up a permanent colony.

Later, both the French and the British established settlements on the island. Control of St. Lucia alternated between the French and the British until the United Kingdom took over in 1814. The island country became independent in 1979.

Mountains and lush forests

Thickly forested mountains cut by deep gorges cover most of St. Lucia. The country's highest peak, Mount Gimie, rises 3,117 feet (950 meters) near the center of the island. St. Lucia also boasts a volcano whose crater can be reached by car. This volcano, with its hot springs and steaming vents that spurt hot gases and sulfur, lies near the spectacular twin peaks of Gros Piton and Petit Piton. The Pitons are two forested volcanic *plugs* (cylindrical masses of rock formed in the craters of ancient volcanoes).

Lush tropical vegetation covers large areas of the island. Bougainvillea, hibiscus, orchids, and roses grow abundantly in the rain forest, which is also home to many beautiful Caribbean birds.

Way of life

Although the people of nearby islands have either a distinct British or French character, St. Lucians enjoy a charming combination of the two, along with an African influence. For example, English is the official language, but many people speak a French dialect. Most of the geographical names on maps of St. Lucia are also in French. The majority of the people

FACTS

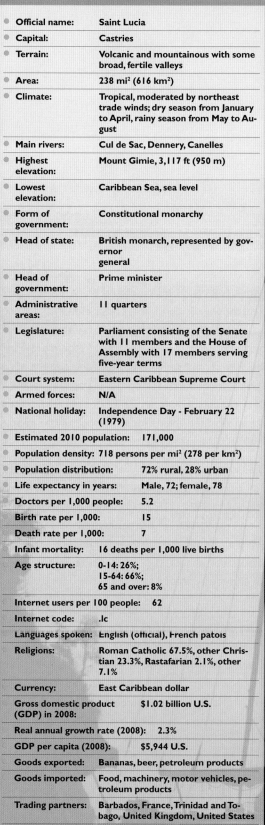

Official name:	Saint Lucia
Capital:	Castries
Terrain:	Volcanic and mountainous with some broad, fertile valleys
Area:	238 mi² (616 km²)
Climate:	Tropical, moderated by northeast trade winds; dry season from January to April, rainy season from May to August
Main rivers:	Cul de Sac, Dennery, Canelles
Highest elevation:	Mount Gimie, 3,117 ft (950 m)
Lowest elevation:	Caribbean Sea, sea level
Form of government:	Constitutional monarchy
Head of state:	British monarch, represented by governor general
Head of government:	Prime minister
Administrative areas:	11 quarters
Legislature:	Parliament consisting of the Senate with 11 members and the House of Assembly with 17 members serving five-year terms
Court system:	Eastern Caribbean Supreme Court
Armed forces:	N/A
National holiday:	Independence Day - February 22 (1979)
Estimated 2010 population:	171,000
Population density:	718 persons per mi² (278 per km²)
Population distribution:	72% rural, 28% urban
Life expectancy in years:	Male, 72; female, 78
Doctors per 1,000 people:	5.2
Birth rate per 1,000:	15
Death rate per 1,000:	7
Infant mortality:	16 deaths per 1,000 live births
Age structure:	0-14: 26%; 15-64: 66%; 65 and over: 8%
Internet users per 100 people:	62
Internet code:	.lc
Languages spoken:	English (official), French patois
Religions:	Roman Catholic 67.5%, other Christian 23.3%, Rastafarian 2.1%, other 7.1%
Currency:	East Caribbean dollar
Gross domestic product (GDP) in 2008:	$1.02 billion U.S.
Real annual growth rate (2008):	2.3%
GDP per capita (2008):	$5,944 U.S.
Goods exported:	Bananas, beer, petroleum products
Goods imported:	Food, machinery, motor vehicles, petroleum products
Trading partners:	Barbados, France, Trinidad and Tobago, United Kingdom, United States

St. Lucia, an independent nation within the British Commonwealth, is situated about 240 miles (385 kilometers) north of Venezuela. Almost three-quarters of the islanders live in rural areas.

A shipment of cocoa beans triggers some spirited bargaining among a group of St. Lucian traders. Most of the island's people are the descendants of African slaves who were brought in to work on the plantations during the days of British and French colonial rule.

On the southwest coast of St. Lucia, Petit Piton—with its famous sugar-loaf shape—overlooks a small St. Lucian village. Like Gros Piton, its twin peak, Petit Piton is of volcanic origin. On the northwest coast stands the capital city of Castries, an attractive town with one of the most beautiful natural harbors in the Caribbean. Visitors strolling through Castries will find large Victorian gardens—a reminder of the island's British heritage.

are Roman Catholics—another sign of French influence. But red telephone boxes, identical to those used in the United Kingdom, can be found throughout the island.

St. Lucia's economy is based on agriculture, and the islanders use most of the produce they grow. Few crops, among them bananas and coconuts, are exported. Industry plays a minor role in the economy, with only a small number of factories manufacturing clothing, electrical parts, paper products, and textiles.

A good road network and modern hotels have helped build St. Lucia's fast-growing tourist industry. Visitors to the island enjoy the superb beaches along the west coast, as well as the awe-inspiring mountain scenery and the delightful French-Creole cuisine. Tourists who enjoy exploring the countryside will find charming villages, ruined forts, lighthouses—and even traces of pirate hideaways.

SAINT VINCENT AND THE GRENADINES

St. Vincent and the Grenadines is a small island nation in the Lesser Antilles. The country consists of the island of St. Vincent and about 100 small islands of the Grenadine chain.

Arawak Indians were the first inhabitants of what is now St. Vincent and the Grenadines. They were conquered by Carib Indians about 1300. Christopher Columbus became the first European to arrive in St. Vincent. The Carib, British, and French fought for control of the island until 1783, when the British gained control. Most of the Carib refugees were deported to the island of Roatán, now part of Honduras.

During the 1900's, St. Vincent and the Grenadines gradually gained more freedom from the United Kingdom. The islands became an independent nation in 1979.

Volcanic islands

The islands of St. Vincent and the Grenadines were formed by volcanic eruptions, and tropical vegetation covers much of the land. The northern part of St. Vincent is dominated by an active volcano—Mount Soufrière—which rises 4,048 feet (1,234 meters) above sea level. When it last erupted in April 1979, rivers of lava destroyed large areas of crops. About 20 percent of the island's people were forced to evacuate their homes.

Most of the Grenadines are hilly and wooded. Along the coasts, coral reefs enclose many fine beaches of white or volcanic black sand. The sparkling blue waters surrounding the Grenadines are among the finest sailing areas in the Caribbean Sea.

Kingstown, the nation's capital and main port, lies on the southwest coast of St. Vincent. The city's famous Botanical Gardens, said to be the

Wood and concrete houses with corrugated metal and tile roofs line the streets of Kingstown, the capital and largest city of St. Vincent and the Grenadines. Kingstown's beautiful harbor on the southwest coast of St. Vincent Island is a center for import and export activities.

FACTS

Official name:	Saint Vincent and the Grenadines
Capital:	Kingstown
Terrain:	Volcanic, mountainous
Area:	150 mi² (389 km²)
Climate:	Tropical; little seasonal temperature variation; rainy season (May to November)
Main rivers:	Colonarie, Cumberland
Highest elevation:	Mount Soufrière, 4,048 ft (1,234 m)
Lowest elevation:	Caribbean Sea, sea level
Form of government:	Constitutional monarchy
Head of state:	British monarch, represented by governor general
Head of government:	Prime minister
Administrative areas:	6 parishes
Legislature:	House of Assembly with 21 members serving five-year terms
Court system:	Eastern Caribbean Supreme Court
Armed forces:	N/A
National holiday:	Independence Day - October 27 (1979)
Estimated 2010 population:	122,000
Population density:	813 persons per mi² (314 per km²)

The peaceful island of Mustique has become a fashionable destination for wealthy Europeans. Many have built vacation homes on the island.

The island nation of St. Vincent and the Grenadines lies in the Caribbean Sea about 200 miles (320 kilometers) north of Venezuela.

Population distribution:	56% rural, 44% urban
Life expectancy in years:	Male, 71; female, 75
Doctors per 1,000 people:	0.8
Birth rate per 1,000:	16
Death rate per 1,000:	7
Infant mortality:	17 deaths per 1,000 live births
Age structure:	0-14: 28%; 15-64: 65%; 65 and over: 7%
Internet users per 100 people:	60
Internet code:	.vc
Languages spoken:	English (official), French patois
Religions:	Anglican 47%, Methodist 28%, Roman Catholic 13%, other 12%
Currency:	East Caribbean dollar
Gross domestic product (GDP) in 2008:	$597 million U.S.
Real annual growth rate (2008):	5.0%
GDP per capita (2008):	$4,937 U.S.
Goods exported:	Arrowroot, bananas, rice
Goods imported:	Food, machinery, petroleum products, transportation equipment
Trading partners:	Barbados, France, Trinidad and Tobago, United States

oldest in the West Indies, contain a breadfruit tree brought from Tahiti in 1792 by William Bligh, a famous British sea captain. Bligh intended the breadfruit tree to be a source of food for the African slaves who had been brought in to work on the plantations. Breadfruit later became a staple food of the West Indies.

People

Today, most of the people of St. Vincent and the Grenadines are of African descent. The islands are also home to a number of Asians whose ancestors came to the islands as contract laborers. English is the official language, but many islanders also speak French.

The economy of St. Vincent and the Grenadines is based on agriculture and tourism. The islands' rich volcanic soil is ideal for growing a variety of tropical crops, such as bananas and coconuts. The country leads the world in the production of arrowroot, a plant whose roots are made into starch. Most tourism in the country occurs in the Grenadines.

SAMOA

Samoa is an independent island country in the Pacific Ocean. It lies in the western part of the Samoa Islands chain, about 1,700 miles (2,740 kilometers) northeast of New Zealand. American Samoa, a United States territory, occupies the eastern part of the Samoa chain.

Samoa is one of the smallest countries in the world. It consists of two main islands, Upolu and Savai'i, and several smaller islands. The islands were formed by erupting volcanoes. Tropical rain forests cover the high volcanic peaks at the center of the islands, while the shores are lined with tall, graceful coconut palms and fringed with coral reefs.

Way of life

Most Samoans are Polynesian and live with their relatives in extended family groups called *aiga*. The aiga elects a head of the family, called a *matai*.

The people speak Samoan, a Polynesian language. Many Samoans also speak English. Most of the people can read and write Samoan, and about 50 percent can read and write English. Almost all Samoans are Christians.

Many Samoans live in open-sided houses that have a thatched roof supported by poles. Most Samoan men wear a *lava-lava,* a piece of cloth that is wrapped around the waist like a skirt and sometimes worn with a blouse or shirt. Most of the women wear a long lava-lava and an upper garment called a *puletasi.*

Samoa's economy is based on agriculture, and about 70 percent of the people are farmers. The chief food crops are bananas, coconuts, breadfruit, and *taro* (a starchy tuber). The people also raise pigs and chickens and catch fish for food. They raise most of their own food, build their own houses, and make most of their own clothing.

History and government

People have lived in Samoa for at least 2,000 years, probably migrating there from what are now Fiji and Vanuatu. The Samoans began forming their own nation about 1,000 years ago. Many chiefs ruled the people until a woman named Salamasina united them in the 1500's.

FACTS

Official name:	Independent State of Samoa
Capital:	Apia
Terrain:	Narrow coastal plain with volcanic, rocky, rugged mountains in interior
Area:	1,093 mi² (2,831 km²)
Climate:	Tropical; rainy season (October to March), dry season (May to October)
Main rivers:	N/A
Highest elevation:	Mount Silisili (on Savai'i), 6,095 ft (1,858 m)
Lowest elevation:	Pacific Ocean, sea level
Form of government:	Mix of parliamentary democracy and constitutional monarchy
Head of state:	Chief
Head of government:	Prime minister
Administrative areas:	11 districts
Legislature:	Fono (Legislative Assembly) with 49 members serving five-year terms
Court system:	Supreme Court, Court of Appeal
Armed forces:	New Zealand is responsible for Samoa's defense
National holiday:	Independence Day Celebration June 1 (1962)
Estimated 2010 population:	192,000
Population density:	176 persons per mi² (68 per km²)
Population distribution:	77% rural, 23% urban
Life expectancy in years:	Male, 71; female, 74
Doctors per 1,000 people:	0.3
Birth rate per 1,000:	28
Death rate per 1,000:	6
Infant mortality:	22 deaths per 1,000 live births
Age structure:	0-14: 39%; 15-64: 56%; 65 and over: 5%
Internet users per 100 people:	5
Internet code:	.ws
Languages spoken:	Samoan (Polynesian), English
Religions:	Congregationalist 34.8%, Roman Catholic 19.6%, Methodist 15%, Latter-Day Saints 12.7%, Assembly of God 6.6%, other Christian 9.3%, other 2.0%
Currency:	Tala
Gross domestic product (GDP) in 2008:	$537 million U.S.
Real annual growth rate (2008):	3.3%
GDP per capita (2008):	$2,872 U.S.
Goods exported:	Agricultural products, beer, fish products
Goods imported:	Food, machinery
Trading partners:	Australia, Fiji, New Zealand, United States

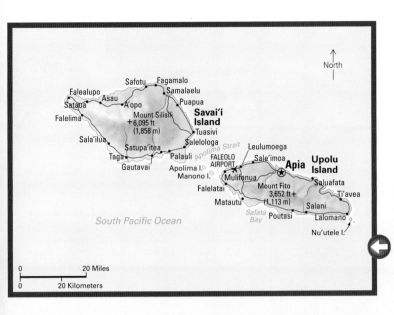

Samoa, one of the smallest countries in the world, lies about 1,700 miles (2,740 kilometers) northeast of New Zealand. The nation consists of two main islands—Upolu and Savai'i—and several smaller islands.

A Samoan matai, the chief of an extended family, is elected by the family group. The chiefs, in turn, elect the members of the country's parliament.

Rich vegetation covers the fertile land near a coastal inlet. Farther inland, tropical rain forests blanket high volcanic peaks.

A beach in Samoa provides a scenic location for a kindergarten class. Education is free, and elementary school is compulsory. Most villages have government-run elementary schools.

The first European to visit the islands, a Dutch explorer named Jacob Roggeveen, arrived in 1722. But few others followed until the first mission was established in Savai'i in 1830. During the mid-1800's, two royal families ruled different parts of Samoa and fought over who would be king. Germany, the United Kingdom, and the United States supported rival groups. In 1899, the three countries agreed that Germany and the United States would divide the islands and, in 1900, Germany took control of the western islands.

During World War I (1914-1918), New Zealand occupied the islands. The New Zealand government became increasingly unpopular, and eventually Samoans refused to obey laws or cooperate with the government. After World War II (1939-1945), the United Nations made what was then called Western Samoa a trust territory and asked New Zealand to prepare the islands for independence. Western Samoa became independent in 1962. It joined the United Nations in 1976. It changed its name to Samoa in 1997.

The country has a 49-member parliament, called the Fono, that elects the prime minister and passes laws. The prime minister runs the government with a cabinet selected from among the Fono members. Laws passed by the Fono do not go into effect until the head of state approves them.

SAN MARINO

San Marino is a tiny independent state surrounded entirely by Italy. It lies in the northeastern foothills of the Apennine Mountains. San Marino covers only 24 square miles (61 square kilometers), making it the third smallest country in Europe, after Vatican City and Monaco.

The capital city, also called San Marino, lies on the upper slopes of Mount Titano, the country's highest mountain. Mount Titano has three peaks, each crowned by a tower built during the Middle Ages and linked by a huge stone wall. The towers are re-created on San Marino's postage stamps, which are prized by collectors all over the world.

Beginnings

Although it is completely surrounded by Italy, San Marino has a long tradition of independence. The country's ability to remain independent throughout the 1800's, when the surrounding Italian city-states were being absorbed by European rulers, is evidence of the proud spirit of its people.

Legend has it that San Marino was founded by, and named after, Saint Marinus, a Christian stonecutter. During the A.D. 300's, Marinus fled to Mount Titano to escape religious persecution by the Romans.

Over the centuries, an independent religious community developed in the area. By the 1300's, San Marino had become a republic. In 1631, the pope recognized San Marino's independence.

A thriving economy

Most of San Marino's income comes from the tourist trade. Visitors come to this country to enjoy the breathtaking mountaintop views and the sunny, mild climate. Ceramics, jewelry, leather goods, and silks made by San Marinese craft workers fill the souvenir shops.

Farming is also important to the San Marinese economy. Although the soil in the countryside is rocky and poor, the mild climate and frequent rainfall enable farmers to grow corn, grapes for wine, olives, and wheat. Herds of cattle and sheep that graze on the mountain slopes provide milk, cheese, butter, wool, and hides. Construction materials are also important to the econ-

FACTS

Official name:	**Repubblica di San Marino (Republic of San Marino)**
Capital:	**San Marino**
Terrain:	**Rugged mountains**
Area:	**24 mi² (61 km²)**
Climate:	**Mediterranean; mild to cool winters; warm, sunny summers**
Main rivers:	**Ausa, Fiumicello, San Marino, Marano**
Highest elevation:	**Mount Titano, 2,478 ft (755 m)**
Lowest elevation:	**Ausa River at northern border, 164 ft (50 m)**
Form of government:	**Republic**
Head of state:	**Two captains-regent**
Head of government:	**Secretary of state for foreign and political affairs**
Administrative areas:	**9 castelli (municipalities)**
Legislature:	**Consiglio Grande e Generale (Grand and General Council) with 60 members serving five-year terms**
Court system:	**Consiglio dei XII (Council of Twelve)**
Armed forces:	**Italy is responsible for San Marino's defense**
National holiday:	**Founding of San Marino - Sept. 3, 301**
Estimated 2010 population:	**32,000**
Population density:	**1,333 persons per mi² (525 per km²)**
Population distribution:	**89% urban, 11% rural**
Life expectancy in years:	**Male, 79; female, 85**
Doctors per 1,000 people:	**47.4**
Birth rate per 1,000:	**10**
Death rate per 1,000:	**8**
Infant mortality:	**4 deaths per 1,000 live births**
Age structure:	**0-14: 16%; 15-64: 67%; 65 and over: 17%**
Internet users per 100 people:	**51**
Internet code:	**.sm**
Language spoken:	**Italian**
Religion:	**Roman Catholic**
Currency:	**Euro**
Gross domestic product (GDP) in 2008:	**$1.70 billion U.S.**
Real annual growth rate (2007):	**4.3%**
GDP per capita (2008):	**$58,724 U.S.**
Goods exported:	**N/A**
Goods imported:	**N/A**
Trading partners:	**N/A**

The independent state of San Marino is completely surrounded by Italy. A highway links the country with Rimini, the nearest Italian city.

omy. Mountainside quarries yield large amounts of lime and building stone, as well as the raw materials for a busy tile industry.

San Marino has a close economic relationship to Italy. Their trading agreement dates from 1862, when Italy guaranteed San Marino protection and financial assistance in return for tax benefits and a monopoly on San Marino's salt and tobacco imports.

A proud and happy people

The San Marinese are closely linked to the Italian culture too, though they prize their independence. Like the Italians, the San Marinese speak Italian, and most of them are Roman Catholics.

The San Marinese enjoy a high standard of living. Nearly everyone in San Marino can read and write. The law requires children from the ages of 6 to 14 to attend school.

Spectacular mountaintop views, as well as mild, sunny weather, draw millions of visitors to San Marino each year.

The walled city of San Marino atop Mount Titano is the capital of tiny San Marino, an independent state since the A.D. 300's. San Marino lies about 12 miles (19 kilometers) from the Adriatic Sea.

SÃO TOMÉ AND PRÍNCIPE

The small African country of São Tomé and Príncipe consists of two main islands in the Gulf of Guinea—São Tomé Island and Príncipe Island—and several tiny ones. Although fairly small itself, São Tomé Island is much larger than Príncipe Island and is home to almost 95 percent of the country's total population.

The islands are part of a series of extinct volcanoes. On both of the main islands, the western end rises sharply from the sea. Inland, steep basalt rock formations rise toward the center. The land slopes gradually downward to the eastern shore, where volcanic ash has created rich and fertile soil.

Partly because of these soil deposits, the economy of São Tomé and Príncipe is based on agriculture, though fishing is also important. Most of the nation's cultivated land is owned and operated by large commercial farms, while the remainder is divided among thousands of small farm owners. The chief agricultural products are bananas, cocoa, coconuts, coffee, copra, and livestock.

History

The history of the islands is closely connected with agriculture. The islands were uninhabited when Portuguese explorers discovered them in 1470. Around 1485, Portugal began to send convicts, exiles, and settlers to the islands. These people tried to grow sugar cane, but they needed more workers to produce large crops. The Portuguese then started importing slaves from the African mainland to work on the plantations. The islands soon ranked among the world's leading producers of sugar.

In the mid-1500's, slaves on São Tomé Island revolted, and some plantation owners abandoned their estates. Sugar production declined, but another economic activity soon replaced it—the slave trade. Slaves from mainland Africa were sent to São Tomé before being shipped to the Americas and elsewhere.

FACTS

Official name:	Republica Democratica de São Tomé e Príncipe (Democratic Republic of São Tomé and Príncipe)
Capital:	São Tomé
Terrain:	Volcanic, mountainous
Area:	372 mi² (964 km²)
Climate:	Tropical; hot, humid; rainy season (October to May)
Main rivers:	N/A
Highest elevation:	Pico de São Tomé, 6,640 ft (2,024 m)
Lowest elevation:	Atlantic Ocean, sea level
Form of government:	Republic
Head of state:	President
Head of government:	Prime minister
Administrative areas:	2 provinces
Legislature:	Assembleia Nacional (National Assembly) with 55 members serving four-year terms
Court system:	Supreme Court
Armed forces:	N/A
National holiday:	Independence Day - July 12 (1975)
Estimated 2010 population:	166,000
Population density:	446 persons per mi² (172 per km²)
Population distribution:	60% urban, 40% rural
Life expectancy in years:	Male, 65; female, 68
Doctors per 1,000 people:	0.5
Birth rate per 1,000:	37
Death rate per 1,000:	7
Infant mortality:	57 deaths per 1,000 live births
Age structure:	0-14: 44%; 15-64: 52%; 65 and over: 4%
Internet users per 100 people:	15
Internet code:	.st
Language spoken:	Portuguese (official)
Religions:	Catholic 70%, other Christian 10%, other 20%
Currency:	Dobra
Gross domestic product (GDP) in 2008:	$175 million U.S.
Real annual growth rate (2008):	5.5%
GDP per capita (2008):	$1,036 U.S.
Goods exported:	Mostly: cocoa, also: copra
Goods imported:	Food and beverages, petroleum products, transportation equipment
Trading partners:	Netherlands, Portugal, United States

A small church in the town of Pantufo, near the capital city of São Tomé, attracts worshipers. Roman Catholicism is the main religion among the Europeans and Creoles in the country.

A palm-fringed beach lines one of the islands of São Tomé and Príncipe.

São Tomé and Príncipe gets its name from its two main islands. The country's capital city, also called São Tomé, sits on the island of São Tomé. The islands lie in the Gulf of Guinea, about 180 miles (290 kilometers) from the mainland.

In the 1800's, Portuguese planters began to grow coffee and cacao on the islands, using slave labor. When slavery ended later in the 1800's, the planters imported contract laborers from mainland Africa. These laborers were treated harshly, and from time to time they tried to revolt. As recently as 1953, Portuguese troops killed hundreds of such protesting workers in an incident known as the Batepa massacre.

Also in the mid-1900's, many people in São Tomé and Príncipe began demanding an end to Portuguese rule. They won their independence on July 12, 1975.

Modern life

Today, about 70 percent of the country's people are of mixed African and European descent. They are sometimes called *Creoles*. Many own small farms or businesses; others work as fishermen or laborers. People from the island country of Cape Verde and mainland African countries form the second largest group on the islands. These people generally work as laborers at low-paying jobs. Europeans, a small percentage of the population, generally own farms or work at technical or management jobs.

Many of the Creoles and Europeans speak an old form of Portuguese and practice Roman Catholicism. Africans from Cape Verde and the mainland generally use the language and follow the religion of their place of birth.

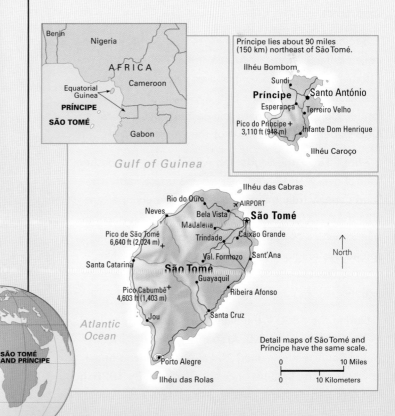

Saudi Arabia occupies most of the Arabian Peninsula, stretching from the Red Sea in the west to the Persian Gulf in the east. The kingdom of Saudi Arabia has a long history. Its royal family goes back more than 500 years, and Islam, one of the world's great religions, began in what is now Saudi Arabia.

Several thousand years ago, various Semitic peoples lived in the area. The Sabaeans were prosperous traders in the southwestern part of the peninsula, the Nabataeans controlled caravan trade routes in the northwestern part, and Arab nomads called Bedouins roamed the interior.

About A.D. 570, Muhammad—the founder of Islam— was born in the city of Mecca. At that time, when most Arabians believed in many gods, Muhammad preached the belief in one God, Allah. Many Meccans opposed Muhammad, and he fled to Medina in 622. But he re- turned to Mecca with an army eight years later and cap- tured the city. He converted the people to his religion, which he called Islam. When Mohammad died in 632, much of Arabia was under the rule of *Muslims,* followers of Islam.

During the mid-600's, Muhammad's successors, called *caliphs,* conquered lands to the east, west, and north of Arabia. But when the Muslim capital was moved north to Syria, the Arabian Peninsula declined in importance.

Beginning about 750, numerous warring groups con- trolled different areas of Arabia. In the early 1500's, the Ottoman Turks gained control over parts of western Arabia. In the 1800's, the British established protec- torates along the southern and eastern coasts. Local Arab leaders still ruled most of the interior.

About 1500, the Sauds—an Arab family—established control over a small area near what is now Riyadh. The Saud *dynasty* (ruling family) was relatively unimportant until the 1700's, when Muhammad ibn Saud formed an alliance with a Muslim religious reformer. Saudi armies spread this religious movement over most of Arabia, and the Saud family took control of the converted regions.

During the 1800's, the Sauds lost much of their land to the Ottomans and rival Arab tribes. But in 1902, a young Saudi leader named Abd al-Aziz ibn Saud began to win back the land his ancestors had lost. By 1925, he had captured western, central, and eastern Arabia, and in 1932 he united these regions as the Kingdom of Saudi Arabia.

SAUDI ARABIA

The new kingdom was poor and undeveloped, but in 1933 Ibn Saud granted an American oil company the right to explore for and produce oil in Saudi Arabia. A major oil deposit was discovered in 1938, and large-scale production began in 1945.

The oil industry developed rapidly and brought incredible wealth to Saudi Arabia. Ibn Saud used the oil profits to build schools, hospitals, and roads. When Ibn Saud died in 1953, his oldest son, Saud, became king, and his younger son, Faisal, became crown prince and prime minister.

Saud was forced off the throne in 1964, and Faisal became king. King Faisal continued to modernize Saudi Arabia, building more hospitals, schools, and apartment buildings. The country's transportation and communications systems were improved and expanded, television was introduced, and Saudi Arabia began to take a more active role in Arab and world affairs.

In the 1970's, the nation used its oil to influence international politics. Saudi Arabia temporarily cut off oil exports to the United States and the Netherlands when those two countries supported Israel in an Arab-Israeli war. Along with other oil-producing countries, it raised prices sharply, greatly increasing its income.

In March 1975, King Faisal was assassinated by his nephew. But his successors—Khalid, who reigned until his death in 1982, and Fahd, who succeeded Khalid—generally continued Faisal's policies, and Saudi Arabia continued to play an important role in Arab and world politics.

In 1990, Iraqi forces invaded and occupied Kuwait. Many people feared that Saudi Arabia would be Iraq's next target, and the United States and other countries sent troops to help defend Saudi Arabia at King Fahd's request. War broke out in January 1991, and the coalition forces quickly defeated Iraq.

King Fahd died in 2005, and his half brother Abdullah became king and prime minister. Also in 2005, municipal elections were held throughout Saudi Arabia for the first time ever. They were the country's first political elections of any kind since 1963.

SAUDI ARABIA TODAY

Saudi Arabia is a large Middle Eastern nation in which oil wealth has created a land of change and contrast. Cars and trucks speed through areas where once only camels traveled. High-rise apartment buildings are replacing mud houses on city streets.

Saudi Arabia is a kingdom based on the laws of Islam. The king is both the chief political leader and the supreme religious leader, or *imam*. Islamic law—the *Sharī`ah*—regulates most of the country's affairs.

The Saudi royal family, which consists of several thousand people, is the most important political group in the nation. Its leading members select the king from among themselves, but their choice must be approved by the *ulema*, the Muslim religious leaders.

The king, who holds both executive and legislative powers, issues royal decrees in matters not covered by the Sharī`ah, such as traffic laws. To assist him, the king appoints a Council of Ministers. Council members advise the king, write the laws, and head government departments. The king serves as prime minister and usually appoints close relatives to important council positions. Saudi Arabia also has a Consultative Council that consists of a chairman and 150 other members, who are all appointed by the king. The council advises the monarch and can propose legislation. The monarch makes final decisions on all changes to Saudi law.

Religious courts handle all civil and criminal cases, with judgments based on the Sharī`ah. There are no juries. The king himself represents the highest court of appeal. In cases involving laws decreed by the king, a Board of Grievances is the final court of appeal. However, the board is responsible to the king.

The only non-Muslims in Saudi Arabia are foreigners. More than 90 percent of the Saudi people belong to the Sunni branch of Islam; the rest are Shiites.

Islam influences many aspects of life in the country, including family relationships, education, and diet. A Saudi father is the head of the family. His wife runs the household, but Saudi women traditionally have had little free-

FACTS

Official name:	Al-Mamlaka Al-Arabiyya Al-Saudiyya (Kingdom of Saudi Arabia)
Capital:	Riyadh
Terrain:	Mostly uninhabited, sandy desert
Area:	830,000 mi² (2,149,690 km²)
Climate:	Harsh, dry desert with great extremes of temperature
Main rivers:	N/A
Highest elevation:	Jabal Sawda, 10,279 ft (3,133 m)
Lowest elevation:	Persian Gulf, sea level
Form of government:	Monarchy
Head of state:	King and prime minister
Head of government:	King and prime minister
Administrative areas:	13 mintaqat (provinces)
Legislature:	Consultative Council with 150 members and a chairman serving four-year term
Court system:	Supreme Council of Justice
Armed forces:	221,500 troops
National holiday:	Unification of the Kingdom - September 23 (1932)
Estimated 2010 population:	26,551,000
Population density:	32 persons per mi² (12 per km²)
Population distribution:	81% urban, 19% rural
Life expectancy in years:	Male, 74; female, 78
Doctors per 1,000 people:	1.4
Birth rate per 1,000:	29
Death rate per 1,000:	3
Infant mortality:	16 deaths per 1,000 live births
Age structure:	0-14: 38%; 15-64: 59%; 65 and over: 3%
Internet users per 100 people:	29
Internet code:	.sa
Language spoken:	Arabic (official)
Religion:	Muslim
Currency:	Saudi riyal
Gross domestic product (GDP) in 2008:	$467.70 billion U.S.
Real annual growth rate (2008):	4.2%
GDP per capita (2008):	$17,741 U.S.
Goods exported:	Mostly: petroleum and petroleum products Also: chemicals, plastics
Goods imported:	Chemicals, food, iron and steel, machinery, motor vehicles
Trading partners:	China, Japan, United States

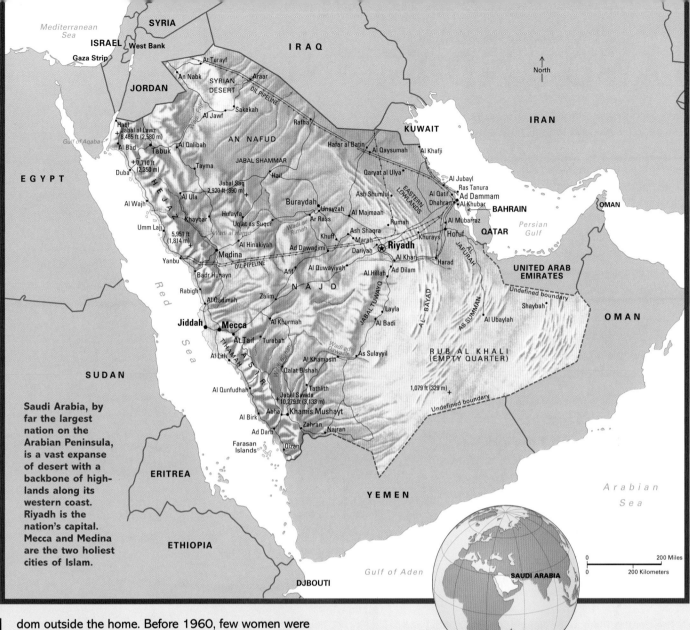

Saudi Arabia, by far the largest nation on the Arabian Peninsula, is a vast expanse of desert with a backbone of highlands along its western coast. Riyadh is the nation's capital. Mecca and Medina are the two holiest cities of Islam.

dom outside the home. Before 1960, few women were educated. In November 1990, about 70 veiled Saudi women gathered in the capital, Riyadh, and drove in a convoy of cars to protest the country's ban on female drivers. The women were stopped and detained by the Islamic police.

The development of the oil industry and the wealth it brought changed Saudi life dramatically. The government used oil profits to improve housing and to provide electricity and other modern conveniences for the people. New schools and adult education programs improved the nation's literacy rate. Girls now attend school, though girls' schools are separate from boys' schools. Increasing numbers of Saudi women work outside the home in teaching, nursing, and small businesses that deal with other women. In 2011, King Abdullah announced that women would be allowed to vote and run for office in elections scheduled for 2015.

Some Muslim religious leaders have voiced strong opposition to these changes in the Saudi way of life. They believe Western influences such as television and the education of women violate the teachings of Islam.

With all the changes, however, much of the traditional Saudi way of life remains. A typical farm village is still a cluster of small stone or mud houses, and the Bedouins still live in tents and roam the deserts as their ancestors did. Most Saudis wear traditional clothing. The men wear a long cotton garment called a *thwab* along with a *ghutra,* a cloth-and-rope-band headpiece that protects them from the sun and wind. Outside the home, most Saudi women cover their face with a veil and wear a floor-length robe called an *abbayah.*

LAND AND ECONOMY

Few people live—and little or nothing grows—in the vast deserts that cover much of Saudi Arabia. But beneath the arid landscapes lie some of the world's largest petroleum deposits.

The land of Saudi Arabia slopes downward from its Western Highlands bordering the Red Sea to its Eastern Lowlands on the Persian Gulf. Between these two mountainous areas lies the vast Central Plateau, bordered on the north and south by deserts.

The northwestern highland region is called Hejaz. The southwestern region is called Asir. In Asir, a narrow coastal plain called the Tihamah separates the rugged mountain peaks from the sea.

Asir is the most fertile area in Saudi Arabia, with annual rainfall of 12 to 20 inches (30 to 51 centimeters). Farmers there grow a variety of crops in the Tihamah and in terraced fields on the mountainsides. Small farm villages lie scattered throughout the region.

The thinly populated northern deserts—An Nafud and the Syrian Desert—and the huge southern desert—the Rub al Khali—get almost no rain. No one lives in the Rub al Khali and only a few groups of Bedouins travel through it.

The Central Plateau, or Najd, gets little rainfall and has little vegetation, but fertile oases in parts of this rocky plateau support small farm communities. The Bedouins bring their animals to graze on the grassy patches that shoot up in the region for a short time after occasional rains.

The oil fields of Saudi Arabia—the world's largest known deposits of petroleum—lie in a sandy, gravelly inland plain in the Eastern Lowlands. The oil industry has led to the development of Dhahran and several other cities and towns in the lowlands. The region also has fertile oases that support large agricultural settlements.

Saudi Arabia's oil fields contain about a fifth of the world's known reserves. The country ranks among the world's largest oil producers, and it is the world's leading oil exporter.

The Ras Tanura oil refinery is located near Saudi Arabia's main oil fields and processes much of its oil. From an offshore dock, tankers load the oil for export.

A comfortable lifestyle, one of the benefits of Saudi Arabia's oil wealth, often includes enjoying afternoon tea with friends. The oil industry has created many jobs, and new towns have developed near the oil fields.

Pipelines at the Red Sea port of Yanbu channel the various grades of oil for export. Saudi oil supplies fuel for automobiles, trucks, and jet aircraft around the world.

Bedouins, like other Saudis, enjoy lively conversation and place great emphasis on providing hospitality to guests. Many of the Bedouin people, who once wandered the deserts, are now farmers or oil workers.

In 1933, Standard Oil of California was given the right to explore for oil in Saudi Arabia. Other oil companies joined Standard and formed the Arabian American Oil Company (Aramco) in 1944. In 1973, the Saudi Arabian government took over part ownership of Aramco's oil facilities. The following year, it began negotiations to take full control. The take-over became official in 1988, when the Saudi Arabian Oil Company (Saudi Aramco) was established.

Modern irrigation methods have improved Saudi agriculture, enabling farmers to produce more crops and a greater variety of fruits and vegetables, including dates, melons, wheat, and tomatoes. Herders and farmers raise cattle, goats, and sheep for dairy products and meat. Chickens and eggs are also produced. Nevertheless, Saudi Arabia still imports much of its food.

The country's small fishing industry produces mostly shrimp. Saudi Arabia has a few manufacturing industries, mainly producing cement, fertilizer, food products, petrochemicals, and steel. Construction has grown since the 1960's, when thousands of Saudis began working on government-sponsored building projects.

The government has also worked to develop other industries, with the help of experts from Europe, Japan, and the United States. In 1974, Saudi Arabia and the United States agreed to cooperate in education, science and technology, agriculture, and industrialization. High spending and a decline in oil prices led to some economic woes in the 1990's. As part of its effort to attract foreign investment and diversify the economy, Saudi Arabia acceded to the World Trade Organization (WTO) in 2005 after many years of negotiations.

BIRTH AND SPREAD OF ISLAM

A lofty minaret towers above a mosque in Yemen. The Arabian Peninsula was the birthplace of Muhammad and of Islam, the religion he founded.

Islam, one of the world's largest religions, traces its beginning to the area that is now Saudi Arabia. Muhammad, the founder of Islam, was an Arab born in Mecca about 570. He was orphaned at an early age, and his uncle eventually became his guardian. The young Muhammad may have joined his uncle on caravans journeying through Arabia to Syria, and he probably attended assemblies and fairs in Mecca. In this way, he may have heard people of different faiths express their beliefs.

At the time, Arabia was a wild, lawless land where desert tribes continually fought each other in bloody wars and the poor suffered terribly. Arabs believed in many gods and prayed to idols and spirits.

One day, when Muhammad was meditating in a cave on Mount Hira, a vision appeared to him. Muhammad believed the vision was the angel Gabriel, who called on him to serve as a prophet and proclaim God's message to the people.

Muhammad later began to preach publicly. He taught that there is only one God, Allah, and that he, Muhammad, was Allah's prophet. He also preached that Allah requires people to make *Islam,* or submission, to him. Muhammad tried to replace the old tribal loyalty with equality and loyalty among all Muslims. He preached against the injustice practiced by the wealthy classes in Mecca.

Muhammad's preaching angered and frightened many Meccans. Some began to hate him, and a powerful leader named Omar persecuted him. Finally, in 622, Muhammad fled north to the city of Medina (then called Yathrib). His flight is called the *Hegira.* The people of Medina welcomed Muhammad, accepted his teachings, and became his followers.

As leader of a religion and a community, Muhammad began to make his message into law. He abolished the worship of idols. He limited *polygyny* (marriage to more than one wife) and divorce. He reformed inheritance laws, regulated slavery, and pro-

The mosque at Isfahan, Iran, is decorated in glowing colors. Most Muslims in Iran belong to the Shiah sect, but Shiites make up less than 20 percent of all Muslims. The largest sect is the Sunni.

Preparations for the pilgrimage to Mecca include a thorough shave. All able Muslims are required to make a pilgrimage, or hajj, to Mecca (see illustration pages 144-145) at least once during their lifetime.

moted care of the poor. Some tribes had avoided the risk of poverty by killing unwanted baby girls; Muhammad banned this practice. He also banned war and violence except in self-defense and to further the cause of Islam.

Muhammad began to attack caravans from Mecca, and Meccans went to war against Muhammad and Medina several times. In 630, Muhammad and his followers returned to Mecca and occupied the city. They destroyed all the idols in the principal Arab shrine and turned it into a *mosque* for Muslim prayer. Meccans then accepted Islam and acknowledged Muhammad as a prophet.

Islam's spread over the Middle East and North Africa began with conquests launched from Mecca and Medina. After Muhammad died in 632, Abu Bakr was elected to rule as *caliph.* He and his successors encouraged the *jihad,* or holy war, to spread Islam. Within 100 years, they built an empire that stretched from northern Spain to India.

This rapid spread engulfed a Persian empire and much of the Christian Byzantine Empire. In the early 700's, the Muslims threatened Western Europe until the Frankish King Charles Martel defeated them at the Battle of Poitiers (or Tours) in 732. This battle may have been one of the most important in history. It determined that Christianity, rather than Islam, would dominate Europe and, eventually, the Americas.

Later, two groups of Turks created their own Muslim empires. The Seljuks were Muslim Turks from central Asia who established an empire in southwestern Asia. The Ottomans were Muslim Turks who succeeded the Seljuks and ruled a vast empire from about 1300 until 1922.

Like many religions, Islam has its sects. As early as the 600's, the Muslim world split into two great divisions, the Sunni and the Shi`ah. The two groups disagreed on leadership after Muhammad died. Sunnis believe that leadership passed to caliphs from Muhammad's tribe upon his death. Shi`ites believe leadership was restricted to descendants of Ali and Fatima, the cousin and daughter of Muhammad. Throughout Islamic history, there has been disagreement and hostility between the two sects.

Muslim conquests in the A.D. 600's and 700's spread Islam into Africa, Europe, and western Asia from its birthplace on the Arabian Peninsula. Many followers of other faiths were converted to Muhammad's new religion.

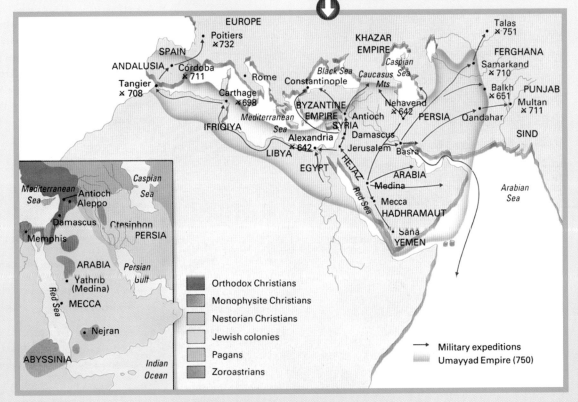

SENEGAL

Senegal is a small country on the huge northwestern bulge of the African continent. The tiny nation of Gambia separates the southern part of Senegal, called the Casamance, from the larger northern part. Dakar, Senegal's capital city, lies along the sandy coast on a peninsula called Cap Vert.

Senegal and Gambia have always had close relations. In 1982, they formed a confederation called Senegambia to strengthen their economic ties and unite their armed forces. However, the confederation was dissolved in 1989 after disputes between the two countries.

In 1960, Senegal was granted its independence from France. Today, it is a republic in which the people elect a president as head of state. The president in turn appoints a prime minister to help run the government. All citizens 18 years old and older may vote.

Senegal's Parliament includes a National Assembly and a Senate. The voters elect members of the National Assembly to serve for five years. The country's Senate consists of both appointed and elected members. Senegal has several political parties, including the Socialist Party of Senegal and the Senegalese Democratic Party.

Since independence, Senegal has kept close relations with France. France provides Senegal with economic assistance and military advisers.

Although conditions have improved since independence, poverty is still widespread in Senegal. In addition, Senegal is plagued with health problems and suffers from a severe shortage of doctors. Water and food contaminated by parasites and other impurities cause much serious illness and even death.

History

The area that is now Senegal probably has been inhabited since prehistoric times. From the A.D. 300's through the 1500's, the eastern sections became part of three great west African empires—the Ghana, the Mali, and the Songhai empires.

FACTS

Official name:	République du Sénégal (Republic of Senegal)
Capital:	Dakar
Terrain:	Generally low, rolling plains rising to foothills in southeast
Area:	75,955 mi² (196,722 km²)
Climate:	Tropical; hot, humid; rainy season (May to November) has strong southeast winds; dry season (December to April) dominated by hot, dry, and dust-laden wind
Main rivers:	Sénégal, Casamance, Gambia
Highest elevation:	1,634 ft (498 m), in the southeast
Lowest elevation:	Atlantic Ocean, sea level
Form of government:	Republic
Head of state:	President
Head of government:	Prime minister
Administrative areas:	14 regions
Legislature:	Parliament consisting of the Senate with 100 members and the Assemblée Nationale (National Assembly) with 150 members serving five-year terms
Court system:	Constitutional Court, Council of State; Cour de Cassation (Court of Final Appeals), Court of Appeals
Armed forces:	13,600 troops
National holiday:	Independence Day - April 4 (1960)
Estimated 2010 population:	13,315,000
Population density:	175 persons per mi² (68 per km²)
Population distribution:	58% rural, 42% urban
Life expectancy in years:	Male, 59; female, 62
Doctors per 1,000 people:	Less than 0.05
Birth rate per 1,000:	37
Death rate per 1,000:	10
Infant mortality:	59 deaths per 1,000 live births
Age structure:	0-14: 42%; 15-64: 54%; 65 and over: 4%
Internet users per 100 people:	8
Internet code:	.sn
Languages spoken:	French (official), Wolof, Pulaar, Jola, Mandinka
Religions:	Muslim 94%, Christian 5% (mostly Roman Catholic), indigenous beliefs 1%
Currency:	Communaute Financiere Africaine franc
Gross domestic product (GDP) in 2008:	$13.49 billion U.S.
Real annual growth rate (2008):	4.7%
GDP per capita (2008):	$1,078 U.S.
Goods exported:	Cement, fish, petroleum products, phosphates
Goods imported:	Machinery, petroleum and petroleum products, rice
Trading partners:	France, Mali, Nigeria

A long-legged crowned crane stands guard over the presidential palace in Dakar. Dakar is Senegal's capital and major seaport and business center. The city has serious housing and unemployment problems due to the large numbers of rural Senegalese who have flocked to Dakar.

The Portuguese came to the region in the 1400's. During the next 400 years, the English, French, and Dutch established trading posts along the coast. They traded such goods as cloth, jewelry, iron bars, weapons, and alcohol for African gold, ivory, millet, and ostrich feathers. Slaves were bought and sold as well.

By the mid-1800's, France had acquired all the coastal trading posts and had conquered African kingdoms in the interior. The entire region became a French colony in 1882, and in 1895, the colony became part of the group of colonies called French West Africa.

Many Senegalese began demanding independence during the 1900's. In 1959, Senegal joined the French Sudan (now Mali) in a federation. The federation was given full independence on June 20, 1960, but Senegal withdrew from the federation two months later and became an independent republic. Léopold Sédar Senghor became the nation's first president.

Since independence

The Socialist Party controlled the government from 1960 to 2000. Prime Minister Abdou Diouf succeeded Senghor as president in 1981. Abdoulaye Wade, leader of the Senegalese Democratic Party, won election as president in 2000 and was reelected in 2007. Senegal adopted a new constitution in 2001 that limits the number of terms a president may serve to two. Although Wade insisted on running for a third term in 2012, he was defeated in the election by former Prime Minister Macky Sall, who had the support of the opposition.

Senegal is a small country on Africa's northwest coast. It is farther west than any other African mainland nation. The tiny nation of Gambia divides the southern part of Senegal from the larger northern part.

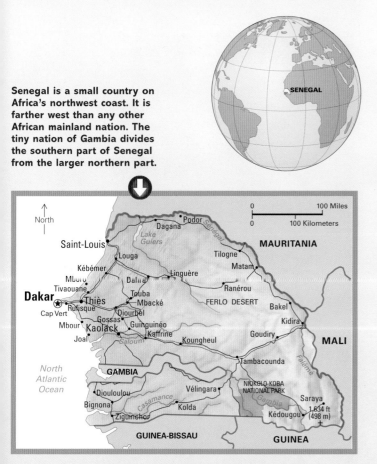

SERBIA

Serbia is a country on the Balkan Peninsula in south-eastern Europe. Serbia is bordered by Hungary on the north; Romania and Bulgaria on the east; Macedonia and Kosovo on the south; and Montenegro, Bosnia-Herzegovina, and Croatia on the west. Serbia includes the province of Vojvodina, in the northern part of the country. Belgrade is the capital and largest city of Serbia. The city lies at the junction of the Danube and Sava rivers and serves as an important port.

Ethnic Serbs make up a majority of the people in Serbia. Other ethnic groups living in the country include Croats, Hungarians, Montenegrins, Romanians, and Slovaks. The population of Serbia is divided almost equally between the cities and rural areas. Many people in the cities work in manufacturing. Agriculture employs most of the people in the countryside. Most of Serbia is mountainous or hilly. However, the Pannonian Plains in the northern part of the country are mainly flat. In general, Serbia has cold, snowy winters and hot summers.

The first united Serbian state was formed in the late 1100's. The Ottoman Empire ruled the region from the mid-1400's to the late 1800's. In 1918, Serbia became part of the Kingdom of the Serbs, Croats, and Slovenes. The kingdom was later renamed Yugoslavia, which means *Land of the South Slavs*. In 1946, Yugoslavia became a Communist federal state made up of six republics: Bosnia-Herzegovina, Croatia, Macedonia, Montenegro, Serbia, and Slovenia. Kosovo and Vojvodina became *autonomous* (self-governing) parts of Serbia, which was the largest and most powerful of the six republics.

The rise of non-Communist political parties in the early 1990's and demands for greater independence by some Yugoslav republics led to the eventual breakup of Yugoslavia. In 1991 and 1992, Slovenia, Croatia, Macedonia, and Bosnia-Herzegovina became independent. Fighting broke out between Serbs and other ethnic groups in Croatia and in Bosnia-Herzegovina. A ceasefire ended most of the fighting in Croatia in January 1992, but some fighting continued. In April 1992, Serbia and Montenegro formed a new, smaller Federal Republic of Yugoslavia.

In 1995, the government of Croatia and the leaders of the Croatian Serbs agreed to end the war in Croatia. Also in 1995, leaders of Bosnia-Herzegovina, Croatia, and Serbia signed a peace treaty. In 1998, fighting began between Serbian forces and ethnic Albanians in Kosovo. Serbian troops withdrew in 1999.

In 2002, the leaders of Serbia, Montenegro, and Yugoslavia developed plans to craft a new constitution and to rename the country. In 2003, Yugoslavia officially became the country of Serbia and Montenegro. In 2006, citizens of Montenegro voted in a referendum to separate from Serbia. Shortly after the referendum, Montenegro declared independence. Serbia then declared its own independence. Kosovo declared its independence from Serbia in 2008.

SERBIA TODAY

Serbia has a republican form of government. A president, elected by voters to a five-year term, serves as the head of state. A one-house parliament called the National Assembly is the country's legislative body. Voters elect the National Assembly's 250 members to four-year terms. A prime minister, who oversees the daily operations of the government, is nominated by the president and elected by the Assembly.

All citizens who are 18 years of age or older may vote. Serbia's major political parties include the Democratic Party, the Democratic Party of Serbia, New Serbia, the Serbian Radical Party, and the Socialist Party of Serbia.

The Supreme Court of Serbia is the country's highest court. Its judges are appointed for life by the National Assembly.

Serbia's flag and coat of arms

Serbia's state flag features red, blue, and white horizontal stripes of equal width. The colors reverse the white-blue-red tricolor of Imperial Russia. The Serbs adopted these "Slavic colors" to honor Russia for its support during Serbia's revolt against the Ottoman Empire in 1804.

Serbia's national coat of arms is set near the *hoist* (end of the flag closest to the staff) of the flag and overlaps the three stripes. The coat of arms has a gold crown above a red shield with a double-headed eagle and two *fleurs-de-lis* (flower of the lily) below the eagle. The fleur-de-lis were added in 1918 to symbolize the country's ruling dynasty.

The coat of arms also shows four Slavic letter *C*'s, which represent four *S*'s in the Slavic system of lettering. Serbians say that the four letters stand for "Only Unity Will Save the Serbs."

After Serbia became part of the Kingdom of the Serbs, Croats, and Slovenes in 1918, it had no flag of its own. Serbia recovered its tricolor flag in 1947, after the establishment of Communist rule in Yugoslavia in 1945. This flag was blue, white, and red and had a large red star outlined in gold in the center. In 1992, the new Federal

FACTS

Official name:	Republic of Serbia
Capital:	Belgrade
Terrain:	In the north, rich fertile plains; in the east, limestone ranges and basins; in the south, mountains and hills
Area:	29,913 mi² (77,474 km²)
Climate:	In the north, cold, snowy winters and hot, humid summers; in the south, cold winters and hot, dry summers
Main rivers:	Danube, Drina, Ibar, Južna Morava, Sava, Timok, Tisa
Highest elevation:	Mount Midžor, 7,113 ft (2,168 m)
Lowest elevation:	115 ft (35 m) above sea level on Danube River at the eastern border with Romania and Bulgaria
Form of government:	Republic
Head of state:	President
Head of government:	Prime minister
Administrative areas:	167 municipalities
Legislature:	National Assembly of the Republic of Serbia with 250 members elected to four-year terms
Court system:	Constitutional Court, Supreme Court of Serbia
Armed forces:	24,300 troops
National holiday:	National Day - February 15
Estimated 2010 population:	7,377,000
Population density:	247 persons per mi² (95 per km²)
Population distribution:	54% urban, 46% rural
Life expectancy in years:	Male, 71; female, 76
Doctors per 1,000 people:	2.0
Birth rate per 1,000:	9
Death rate per 1,000:	14
Infant mortality:	7 deaths per 1,000 live births
Age structure:	0-14: 16%; 15-64: 67%; 65 and over: 17%
Internet users per 100 people:	24
Internet code:	.rs
Languages spoken:	Serbian (official), Hungarian, Bosniak, Romany
Religions:	Serbian Orthodox 85%, Catholic 5.5%, Muslim 3.2%, Protestant 1.1%, other 5.2%
Currency:	Serbian dinar
Gross domestic product (GDP) in 2008:	$50.77 billion U.S.
Real annual growth rate (2008):	5.6%
GDP per capita (2008):	$5,333 U.S.
Goods exported:	Copper products, food, iron and paper products, machinery
Goods imported:	Machinery, petroleum products, transportation equipment
Trading partners:	Bosnia-Herzegovina, Germany, Italy, Russia

SERBIA

defeated the Serbs at the Battle of Kosovo Polje. During Ottoman occupation of Kosovo, many Serbs fled the area. By the 1700's, ethnic Albanians had become a majority. Nevertheless, Kosovo became important in Serbian folklore as the place where Orthodox Christians took a stand against Muslim invaders. Over time, Serbs came to regard Kosovo as their heartland.

Kosovo remained under Ottoman rule after Serbia won its independence in 1878. In 1912, the Serbian army seized the region during the First Balkan War (1912-1913). In 1918, Kosovo became part of the Kingdom of the Serbs, Croats, and Slovenes, later the Kingdom of Yugoslavia.

In 1946, when Yugoslavia became a federal state, Kosovo became an autonomous region, and later an autonomous province, of the republic of Serbia. It had its own parliament and police force. In 1989 and 1990, however, Serbia ended Kosovo's autonomy.

In 1992, ethnic Albanians in Kosovo chose a president and parliament in elections that were declared illegal by the Yugoslav government. In 1997, the Kosovo Liberation Army (KLA), a rebel group seeking independence for Kosovo, attacked Serbian police stations in the province. The next year, Serbian forces attacked Kosovo, killing or evicting many ethnic Albanians.

In 1999, the North Atlantic Treaty Organization (NATO) began air strikes against military targets in Serbia to force the government to accept a peace plan for Kosovo. Serbia withdrew its troops from Kosovo.

Kosovo declared its independence on Feb. 17, 2008. Countries such as France, Italy, the United Kingdom, and the United States recognized Kosovo's independence. However, other countries, such as China, Romania, and Russia, refused to do so.

Republic of Yugoslavia, formed by Serbia and Montenegro, adopted the old Yugoslav flag without the red star.

In 2004, after the country became known as Serbia and Montenegro, the flag reverted to the original red-blue-white tricolor, and the old Serbian coat of arms was added. That design became the Serbian state flag in 2006, after Serbia and Montenegro separated.

Kosovo

Kosovo was a province of Serbia until 2008, when it declared independence. However, the Serbian government has refused to recognize its independence, maintaining that Kosovo is a historic part of Serbia. Most of Kosovo's people are ethnic Albanians.

In the 1100's, Kosovo became part of the first united Serbian state. In 1389, the Ottoman Turks

HISTORY TO 1990

During the A.D. 500's and 600's, groups of Slavs, including the ancestors of modern Serbs, settled in the Balkan Peninsula. Each group had its own leader until the late 1100's, when Stefan Nemanja, a warrior and chief, formed the first united Serbian state.

During the 1300's, Emperor Stefan Dušan led the country in successful wars against the Byzantine Empire. The Serbian empire started to break up after his death in 1355. The Ottoman Empire, based in what is now Turkey, defeated Serbia in the Battle of Kosovo Polje in 1389. The Ottomans gained complete control of Serbia in the mid-1400's and ruled the region for more than 400 years. The Serbs, however, never lost their national pride.

Serbia regained its independence in 1878, following Russia's defeat of the Ottoman Empire in the Russo-Turkish War of 1877-1878. In the First Balkan War (1912-1913), Serbia and the other Balkan states seized almost all Ottoman territory in Europe.

In the early 1900's, political and economic conflicts developed between Serbia and Austria-Hungary. In June 1914, a Serb named Gavrilo Princip assassinated Archduke Franz Ferdinand, the heir to the Austro-Hungarian throne. The assassination touched off World War I, which began a month later when Austria-Hungary declared war on Serbia. After the war ended in 1918, Serbia led the way in forming the Kingdom of the Serbs, Croats, and Slovenes, with Peter I of Serbia becoming its ruler.

Peter's son Alexander succeeded his father in 1921 and later renamed the country *Yugoslavia*. He ruled as a dictator and aggravated tensions among the region's ethnic groups by ignoring their historical borders. Alexander's assassination in 1934 made his 11-year-old son, Peter II, the king. Because of Peter's young age, Alexander's cousin, Prince Paul, governed in the boy's place.

During World War II (1939-1945), Paul's government supported the Axis powers, led by Germany and Italy. However, the Yugoslav army rebelled and overthrew the government. Peter II then took the throne. On April 6, 1941, Germany invaded Yugoslavia. The Yugoslav army surrendered 11 days

TIMELINE

c. 1000 B.C.	Illyrians and Thracians live in what became Yugoslavia.
600's B.C.	Greeks establish colonies along the Adriatic coast.
c. 300's B.C.	Romans invade the Balkan Peninsula.
229-228 B.C. and 219 B.C.	Romans wage war against the Illyrians.
168 B.C.	Romans conquer the Illyrians.
13-9 B.C.	Romans conquer Pannonia.
A.D. 395	The area becomes part of the East Roman (Byzantine) Empire.
1100's	The first united Serbian state is established.
1389	Ottomans defeat Serbia in the Battle of Kosovo Polje.
1400's	Ottomans conquer Serbia.
1804	George Petrovic, a Serbian peasant nicknamed Black George, leads an uprising against the Ottomans.
1815	Miloš Obrenovic leads a second revolt against the Ottomans.
1878	Serbia gains independence from the Ottoman Empire.
1883	Nikola Testa builds first alternating current motor, devising a system that remains the heart of electric power operations worldwide.
1914	Archduke Franz Ferdinand of Austria-Hungary is assassinated by Bosnian Serb Gavrilo Princip, setting off World War I.
1918	The Kingdom of the Serbs, Croats, and Slovenes is formed.
1929	King Alexander I establishes a dictatorship and changes the name of the country to Yugoslavia.
1941	Germany and other Axis troops invade Yugoslavia.
1945	Yugoslavia becomes a Communist state with six republics, led by Josip Broz Tito.
1974	Nine-member Presidency is formed to provide leadership after Tito's retirement or death.
1980	Tito dies.
1989	Slobodan Miloševic becomes president of Serbia.
1990	Yugoslavia's Communist Party votes to end its monopoly on power in the country.
2006	Montenegro separates from Serbia.
2008	Kosovo declares independence.

later. Peter and other government leaders fled to London and formed a government-in-exile.

German and other Axis troops occupied Yugoslavia during most of World War II. A resistance movement against the Axis occupation spread among the Yugoslav people. Some people joined the Partisans, a group led by Josip Broz Tito and the Communist Party. Other Yugoslavs joined the Chetniks, a group led by Draža Mihajlovic. The Partisans wanted to establish a Communist government. The Chetniks supported Peter's government. The two resistance groups fought each other, as well as the occupation forces.

The Partisans gained the support of the Yugoslav people. Aided by Allied troops, the Partisans freed Bel-

**Nicola Tesla
(1856-1943)**

**Alexander I
(1888-1934)**

**Josip Broz Tito
(1892-1980)**

grade from Axis occupation in 1944. The Communist Party then began to govern from the capital. When World War II ended in Europe in May 1945, Tito and the Communists firmly controlled Yugoslavia.

In 1945, Yugoslavia became the Federal People's Republic of Yugoslavia. The monarchy was abolished. The 1946 Constitution organized Yugoslavia into a *federal state*—that is, a state where each republic largely controlled its own affairs. The six republics were Bosnia-Herzegovina, Croatia, Macedonia, Montenegro, Serbia, and Slovenia. Vojvodina and Kosovo became autonomous areas of Serbia. The Communist Party became the only legal political party.

The Communist government took control of farms, factories, and other businesses and worked to develop Yugoslavia from an agricultural country into an industrial one. After a bitter political dispute with the Soviet Union, Tito followed his own style of Communism.

In 1990, Yugoslavia's Communist Party ended its monopoly on power, and each republic held multiparty elections. Non-Communist parties won a majority in the parliaments of Bosnia-Herzegovina, Croatia, Macedonia, and Slovenia. In Serbia and Montenegro, the Communist parties, now known as Socialist parties, retained power.

The monastery of Manasija, built in the early 1400's along the banks of the Resava River, houses splendid examples of Serbian Orthodox painting.

By 2003, Yugoslavia had broken up into five separate countries. Only Serbia and Montenegro remained connected.

Partisan troops march into Belgrade in 1944, liberating the city from Axis occupation. The Partisans, led by Josip Broz Tito and the Communist Party, controlled all of Yugoslavia by the end of World War II.

HISTORY SINCE 1990

In 1989, Slobodan Miloševic, a Serb nationalist, became president of Serbia. His strong belief in Serbian unity helped him gain widespread popular support. His government quickly stripped Kosovo and Vojvodina of their autonomy, and in 1990, it dissolved the government of Kosovo.

In 1991, Croatia, Macedonia, and Slovenia declared their independence from Yugoslavia, following decades of demands for more self-government. Fighting then broke out in Croatia between ethnic Serbs and the Croat militia. The conflict lasted until January 1992. In April 1992, Serbia and Montenegro formed a new Yugoslavia, thus appearing to accept Croatia's independence.

In March 1992, a majority of Bosniaks and ethnic Croats in Bosnia-Herzegovina voted for independence from Yugoslavia in a *referendum* (direct vote) boycotted by ethnic Serbs. Fighting then broke out between Bosnian Serbs and Bosniaks and ethnic Croats.

In late 1995, the Croatian government and the leaders of the Croatian Serbs made peace in Croatia. Representatives of Bosnia, Croatia, and Serbia also agreed to a peace plan for Bosnia-Herzegovina. The plan called for dividing Bosnia into two parts, one to be controlled by a Bosniak-Croat federation and the other by Bosnian Serbs.

In 1997, Yugoslavia's parliament elected Miloševic president of Yugoslavia, though some members boycotted the vote. The next year, Serbian forces attacked villages in Kosovo, killing many people. Miloševic said the attack was a crackdown on the rebel Kosovo Liberation Army, which demanded independence for the province. Fighting began between the Serbian and rebel forces. Serbian forces destroyed villages in the province and drove many of Kosovo's ethnic Albanians from their homes.

In early 1999, the North Atlantic Treaty Organization (NATO) offered a peace plan for Kosovo, but the

TIMELINE

1990	The Yugoslavian government dissolves the government of Kosovo.
1991	Croatia, Macedonia, and Slovenia declare their independence from Yugoslavia; civil war breaks out in Croatia.
1992	Bosnia-Herzegovina declares independence. Serbia and Montenegro form a new, smaller Yugoslavia. Fighting breaks out in Bosnia between Bosnian Serbs and Bosniaks and Croats.
1995	The Croatian government and leaders of the Croatian Serbs agree to a peace plan for Croatia. Bosnia, Croatia, and Serbia agree to a peace plan for Bosnia-Herzegovina.
1996	Yugoslavia establishes diplomatic relations with Bosnia-Herzegovina, Croatia, and Macedonia.
1997	Slobodan Miloševic is elected president of Yugoslavia.
1998	Serbian troops attack Kosovo.
1999	NATO forces attack Yugoslavia to force the government to accept a peace plan for Kosovo; Serbian forces withdraw from Kosovo.
2000	Vojislav Koštunica wins presidential election; Miloševic is ousted from power.
2001	Yugoslav government arrests Miloševic on a variety of corruption charges; he is handed over to the International Criminal Tribunal for the former Yugoslavia to face charges of war crimes.
2002	Miloševic 's war crimes trial begins.
2003	Yugoslavia officially becomes the nation of Serbia and Montenegro; Serbian Prime Minister Zoran Djindjic is assassinated.
2004	Koštunica becomes prime minister of Serbia.
2006	Miloševic dies in prison before his trial ends. Montenegro declares independence; shortly after, Serbia declares its own independence.
2008	Kosovo declares its independence from Serbia; Prime Minister Koštunica resigns.

Slobodan Miloševic (1941-2006), a Serb known for his extreme nationalism, served as president of Yugoslavia from 1997 to 2000. He died while facing a war crimes court for his actions against ethnic Albanians in Kosovo.

Serbian delegates rejected it. In March, NATO began air strikes against military targets in Yugoslavia to force the government to accept the plan. But Serb attacks continued, and hundreds of thousands of people fled Kosovo. In June, Serbia began to withdraw its forces from Kosovo. NATO then stopped the bombing and sent an international peacekeeping force to Kosovo. United Nations (UN) officials served as a temporary regional government, and some refugees returned to Kosovo. However, tensions between ethnic Serbs and Albanians continued to erupt into violence.

Meanwhile, opposition to Miloševic's rule grew. In September 2000, Vojislav Koštunica, the leader of the Democratic Party of Serbia, won Serbia's presidential election. Miloševic and his allies claimed that Koštunica had not won by a large enough margin and called for a runoff election. Protesters demanding Miloševic's resignation soon filled the streets of many of Serbia's major cities, and Miloševic was ousted from power.

In 2001, Miloševic was arrested by the Yugoslav government on corruption charges. After much debate within Yugoslavia and under international pressure, Miloševic was delivered to the International Criminal Tribunal for the former Yugoslavia. He died in prison in 2006, before his trial ended.

In 2003, the parliaments of Yugoslavia, Montenegro, and Serbia created a new constitution and renamed the country Serbia and Montenegro. In 2006, Montenegro declared its independence, shortly after a majority of its citizens voted for separation in a parliament-sponsored referendum. Serbia declared its own independence later that year.

Kosovo declared its independence from Serbia in 2008. Koštunica—who had become prime minister of Serbia in 2004—then resigned because members of his coalition refused to stop pursuing admission to the European Union (EU). Koštunica and his followers wanted to suspend Serbia's application to the EU until all EU member nations acknowledged that Kosovo is part of Serbia. No party won a majority in the election that followed. In June 2008, the Assembly approved a coalition government.

Ethnic Albanians survey damage in Jakovo, a town in Kosovo, after the departure of Serbian forces in June 1999. Jakovo was one of many towns in Kosovo ravaged by fighting between Serbian forces and ethnic Albanians.

Yugoslav protesters rally before the Yugoslav parliament building in Belgrade on Oct. 5, 2000, to demand the resignation of President Slobodan Miloševic. Smoke from burning buildings fills the sky. Miloševic resigned the following day.

ECONOMY

After Josip Broz Tito and his Communist Party came to power in Yugoslavia in 1945, they established two main economic objectives. The first was to transform an economy based on agriculture into one based on industry. The second was to establish what they understood to be the ethic of Communism in the workplace—that is, to place privately owned factories, land, and other economic resources under government control.

When World War II ended in 1945, Yugoslavia was one of the least developed agricultural countries in Europe. About 75 percent of the land held by peasant-farmers consisted of plots less than 12.3 acres (5 hectares) in area. In addition, hundreds of thousands of peasants had no land to work. Those peasants who owned farmland had only basic equipment to work it.

To meet their second economic objective, Yugoslavia's Communist leaders nationalized most businesses and industries. During the 1950's, they set up a system of self-management for these enterprises, in which economic planning was placed in the hands of workers' councils that followed government guidelines.

The Communists successfully transformed Yugoslavia into a modern industrial nation. This rapid industrial growth—which required large imports of capital equipment, fuel, and raw materials—was financed largely by a huge outpouring of credits and grants from the United States and Western Europe. Western countries were willing to aid Yugoslavia because of Tito's foreign policy, which stressed his country's independence from the Soviet Union, the leader of the Communist bloc.

The Yugoslav economy started to decline in the late 1970's, and the country began to experience severe inflation and other economic problems. These difficulties contributed to demands among Yugoslavs for a multiparty political system. In 1990, the Communist government voted to end its monopoly on power. In the multiparty elections that followed, non-Communist candidates won a majority of seats in Bosnia-Herzegovina, Croatia, Macedonia, and Slovenia. In 1992, the new Yugoslav government formed by Socialist parties

A worker assembles a tractor at a factory in Belgrade, a major industrial center. Belgrade's factories also produce automobiles, iron and steel, paper, plastics, and textiles.

Workers smelt gold in a mine in Majdanpek, southeast of Belgrade. In addition to gold, Serbia has deposits of coal, copper, lead, and zinc.

(formerly the Communist parties) in Serbia and Montenegro announced plans to move toward a free enterprise system, in which business owners and managers would decide what to produce and how much to charge for their goods and services.

Agriculture still employs many workers in Serbia. The country's best farmland lies in the province of Vojvodina, in northern Serbia, and in Sumadija, an area south of Belgrade. Serbian farmers grow corn, potatoes, sugar beets, and wheat.

Serbia's rolling hills provide the best areas for livestock breeding. Farmers raise cattle, hogs, and sheep. Also in the hills, farmers tend orchards that produce such fruits as cherries, grapes, plums, and raspberries. Wine is produced in these regions as well.

A farmer harvests wheat near Smederevo. Nearly half of Serbians live in rural areas. Agriculture remains one of the most important sectors of the country's economy.

Manufacturing is a major part of Serbia's economy. Factories in Serbia produce automobiles, cement, iron and steel, plastics, and textiles. Mining is another important industry in Serbia. The country has deposits of coal, copper, lead, and zinc. Wells in the Pannonian Plains of northern Serbia produce petroleum and natural gas.

A network of highways extends from Belgrade, Serbia's capital, but the rest of the country has fewer roads. Roads between some villages are unpaved. Railways link Belgrade with major cities and towns in Serbia and in neighboring countries.

Serbia's rivers, especially the Danube, provide navigable routes for passenger boats and international shipping. Belgrade, which lies at the junction of the Danube and Sava rivers, is a major port city.

Serbia has airports in Belgrade and Niš. The Belgrade airport, which is the country's largest, handles international flights.

PEOPLE

Once a province of ancient Rome, what is now Serbia became part of the East Roman (Byzantine) Empire in A.D. 395. The earliest references to Serbs as a distinct nationality group appear in the writing of the Byzantine Emperor Constantine VII. In his work *De Administrando Imperio,* written in the A.D. 900's, Constantine noted that the Serbs had been converted to Christianity in the 800's.

Serbs make up about 85 percent of the population of what is sometimes called *Serbia proper*—the area excluding the province of Vojvodina. About 55 percent of the people of Vojvodina are Serbs, and nearly 20 percent are Hungarians. The province also includes large numbers of Croats, Montenegrins, Romanians, and Slovaks.

Most of the people of Serbia speak Serbian. The Serbian language traditionally uses a form of the Cyrillic alphabet, the system used in writing Russian and some other Slavic languages. But the Roman alphabet is also used for modern Serbian. Many ethnic groups in Serbia also have their own language.

About half the people of Serbia live in cities. The other half live in rural areas. Most city dwellers live in apartment buildings or older brick houses. Typical suburban housing consists of high-rise apartment buildings made of concrete. Many rural families live in small houses built of brick, stone, or wood with steeply pitched roofs covered in wooden shingles or metal.

The traditional religion of the Serbs is the Eastern Orthodox Church. The Serbian Church played an important role in maintaining the Serbian culture during the Ottoman occupation of Serbia from 1389 to 1878. Serbian Orthodoxy provided a framework for keeping alive a sense of national identity in the face of foreign rule. The wallpaintings that decorate Serbian churches built during the Middle Ages are considered some of the most important examples of early Serbian art.

Some Serbs belong to the Seventh-Day Adventist Church. Ethnic Hungarians and Slovaks typically attend such churches as the Hungarian Evangelical Lutheran Church or the Slovak Evangelical Christian Church.

People enjoy the casual atmosphere of an outdoor cafe in Belgrade, Serbia's capital. The main meal of the day is usually lunch. It often includes soup, followed by a meat or fish dish, a salad, and dessert. A favorite snack is *burek*, a pastry layered with cheese, jam, or meat.

A Serbian Orthodox monk hangs laundry out to dry. The traditional religion of Serbia, the Serbian Orthodox Church, has provided a framework for helping Serbians to maintain their sense of national identity, especially during foreign occupation.

Children in Serbia are required by law to complete at least 8 years of elementary school. However, most children attend school for 12 years. Serbia has universities in Belgrade, Kragujevac, Niš, and Novi Sad.

Most of the adults in Serbia's urban areas are well educated. However, in rural Serbia, some families have traditionally kept girls out of school to work in the home or on the farm. This practice has kept many rural women in domestic roles.

Serbians enjoy many sports, particularly soccer. Basketball is also popular, and almost every town or village in Serbia has its own basketball team.

Serbian folk music is played mainly on the accordion. The violin and the *tamboura,* an instrument resembling a lute, are also used to accompany folk dances in certain parts of Serbia. The best-known traditional dance among the Serbs is the *kolo,* which is performed in a circle.

In the late 1900's, a new, sometimes extreme pride in Serbian culture began to sweep the republic. This movement sparked an interest in Serbian writers of the 1900's, including Matija Bečkovic, Miloš Crnjanski, and Vuk Draškovic.

Serbian cooking reflects both central European and Turkish influences. Serbian cooks are known for grilled, highly seasoned meats and spicy salads. *Cevapcici,* which consists of grilled meatballs served with raw onions on bread, is a Serbian specialty. *Ajvar* is a relish made of roasted red peppers. A favorite snack in Serbia is *burek,* a pastry layered with cheese, jam, or meat. Typical Serbian beverages include thick, sweet Turkish coffee and plum brandy.

A large plant and flower market in Belgrade attracts shoppers.

ENVIRONMENT

Serbia lies in the central part of the Balkan Peninsula in southeastern Europe. The Balkans are a group of countries that cover a peninsula in the southeast corner of Europe. *Balkan* is a Turkish word meaning *mountain*. The Balkans include several mountain ranges.

Most of Serbia is hilly or mountainous. Northern Serbia includes the Pannonian Plains, which are in the Vojvodina province. The plains are a southwestern extension of Hungary's Great Plain. The region is mostly flat, with some low hills. The ancient Pannonian Sea, which once covered this area, disappeared as the surrounding mountains gradually filled it with rich deposits of clay, gravel, sand, and silt. As a result, the Pannonian Plains have the most fertile soil in Serbia, making the region the country's chief agricultural area.

A number of rivers flow through Serbia. They include the Danube, one of Europe's longest waterways. The Danube enters Serbia from Hungary, forming part of the border between Serbia and Croatia. The river flows southeast across Serbia and then forms part of Serbia's border with Romania. The Iron Gate Dam on the Danube stands

Golubac Castle rises on a rocky point high above the Danube River. Built by the Serbs in the early 1300's, it was occupied by the Ottomans in 1391 and retaken by the Serbs in 1867.

Forested mountains surround the Tresnjica Canyon in western Serbia. Most of Serbia's terrain is hilly or mountainous.

at the Iron Gate, a gorge at the border between Serbia and Romania. The power plant of the dam supplies electric power for the two countries.

The Morava River runs north through the hills of southern and central Serbia and then empties into the Danube. The Sava River flows eastward, emptying into the Danube at Belgrade.

The Pannonian Plains have cold winters with a freezing wind called a *košava*. Summers are dry and hot, with temperatures often rising to about 100° F (38° C). In Belgrade, on the edge of the Pannonian Plains, the average January temperature is 32° F (0° C). The average July temperature is 73° F (23° C).

The rest of Serbia has bitterly cold winters with much snow. Heavy rains fall in early summer. Summers are warm in the mountain valleys but cool at higher elevations.

Snow clings to trees in a forest in southern Serbia. This region is hilly or mountainous, with bitterly cold winters and much snow.

BELGRADE

Belgrade is the capital and largest city of Serbia. Belgrade's name in the Serbian language is *Beograd*. The city has been at the center of political and military struggles for thousands of years.

Belgrade, located in north-central Serbia, covers about 71 square miles (184 square kilometers). Belgrade serves as a major river port and railway center. The city lies at the junction of the Danube and Sava rivers. The oldest section of Belgrade, known as Stari Grad (Old City), lies on a hill that overlooks this junction. The most modern section, known as New Belgrade, lies on the west bank of the Sava River.

About 1.6 million people—more than 20 percent of Serbia's population—live in the capital. Most of Belgrade's people are Serbs, but the population also includes ethnic Albanians, Hungarians, Montenegrins, and other groups. The majority of the people use the Serbian language.

Belgrade's way of life combines modern Western influences with Balkan tradition. The city has skyscrapers and other modern buildings, as well as many of the common problems of urban life, including pollution and traffic congestion.

Belgrade has a number of museums, including the Ethnographical Museum and the Fresco Gallery. Theaters in Belgrade present ballets, concerts, operas, and plays. Many parks dot the city, including Kalemegdan, Tashmajdan, and Topchider. The University of Belgrade developed out of a school founded in 1808.

People lived in the vicinity of what is now Belgrade as long ago as 5000 B.C. During the 300's B.C., Celtic tribes settled in the area. The Romans later captured the settlement, which they called Singidunum, and it developed into a city.

In A.D. 441, the Huns destroyed Singidunum. Over the next 1,000 years, invading armies conquered and destroyed the city dozens of times. Slavic tribes captured the city around 600 and later renamed it Beograd. Control of the city shifted frequently between the Hungarians, the Byzantine Empire, and others.

The steeple of the Saborna Crkva (Cathedral Church) rises above a riverfront neighborhood in Belgrade, Serbia's capital and largest city. A major port, Belgrade lies at the junction of the Danube and Sava rivers.

Serbia's parliament building, the seat of the country's National Assembly, sits along one of Belgrade's impressive boulevards.

The Belgrade Fortress, in the oldest part of the city, attracts residents as well as tourists. It offers scenic views of the junction of the Sava and Danube rivers and the surrounding parkland and walking paths.

In 1403, Belgrade became the capital of the Serbian kingdom. The Ottoman Empire captured the city in 1521. As the Ottoman and Austrian empires battled during the 1600's and 1700's, Belgrade changed hands between them several times.

During the 1800's, Belgrade was a center of revolutionary activity by Serbian nationalists fighting for Serbia's independence from Ottoman rule. Serbia won complete independence from the Ottoman Empire in 1878, and Belgrade remained the country's capital.

Austro-Hungarian and German forces occupied Belgrade during World War I (1914-1918). In 1919, the city became the capital of the newly created Kingdom of the Serbs, Croats, and Slovenes, which was later renamed Yugoslavia. The Germans occupied Belgrade during most of World War II (1939-1945). Yugoslav Communists freed Belgrade in 1944.

Belgrade grew rapidly in the mid-1900's. As Yugoslavia broke up in the 1990's, Belgrade experienced much unrest. In 1999, the North Atlantic Treaty Organization (NATO) bombed targets in Belgrade while trying to enforce a peace settlement in Kosovo. In 2000, protests in Belgrade helped lead to the resignation of Yugoslav President Slobodan Miloševic.

SEYCHELLES

The African country of Seychelles consists of about 115 islands scattered over 400,000 square miles (1,000,000 square kilometers) of the Indian Ocean. Some of the islands are composed of granite and some of coral.

The granite islands have white sandy beaches, mountains, streams, and fertile soil, but the land is so rocky that farming is difficult. Some of the coral islands form *atolls* (ring-shaped groupings), while others are low islands with reefs rising only a few feet above the level of the sea. These islands cannot support much plant life, and many are uninhabited.

The great majority of the Seychellois, as the people are called, live on the largest island, Mahé. The vast majority of the people have mixed African, Asian, and European ancestry, and many of the rest are of Chinese or Asian Indian descent.

Most of the people speak Creole, which is a dialect of French. The country's official languages are Creole, English, and French. Most Seychellois are Roman Catholics.

The people's language and religion are a heritage of the islands' colonial past. When the Portuguese discovered the Seychelles in the 1500's, no one lived on the islands. For the next 250 years, the islands served chiefly as a hiding place for pirates.

France claimed the islands in 1756, but a peace treaty gave them to the United Kingdom in 1814. During the early 1970's, many Seychellois began to demand an end to British rule. Their demand was met in June 1976, when Seychelles became independent.

Today, Seychelles is a republic with an elected president as its head. Until 1993, the country's only political party was the Seychelles Peoples United Party. Multiparty elections were held for the first time in July 1993.

Almost all Seychellois children attend elementary school. The polytechnic offers advanced vocational training. Most adults can read and write.

Official name:	Republic of Seychelles
Capital:	Victoria
Terrain:	Some islands are granitic, narrow coastal strip, rocky, hilly; others are coral, flat, elevated reefs
Area:	176 mi² (455 km²)
Climate:	Tropical marine, humid; cooler season during southeast monsoon (late May to September); warmer season during northwest monsoon (March to May)
Main rivers:	N/A
Highest elevation:	Morne Seychellois, 2,993 ft (912 m)
Lowest elevation:	Indian Ocean, sea level
Form of government:	Republic
Head of state:	President
Head of government:	President
Administrative areas:	23 administrative districts
Legislature:	Assemblée Nationale (National Assembly) with 34 members serving five-year terms
Court system:	Court of Appeal, Supreme Court
Armed forces:	200 troops
National holiday:	Constitution Day (National Day) - June 18 (1993)
Estimated 2010 population:	88,000
Population density:	500 persons per mi² (193 per km²)
Population distribution:	54% urban, 46% rural
Life expectancy in years:	Male, 68; female, 77
Doctors per 1,000 people:	1.5
Birth rate per 1,000:	17
Death rate per 1,000:	7
Infant mortality:	12 deaths per 1,000 live births
Age structure:	0-14: 23%; 15-64: 69%; 65 and over: 8%
Internet users per 100 people:	38
Internet code:	.sc
Languages spoken:	Creole, English, French (all official)
Religions:	Roman Catholic 82.3%, Anglican 6.4%, other Christian 4.5%, Hindu 2.1%, other 4.7%
Currency:	Seychellois rupee
Gross domestic product (GDP) in 2008:	$833 million U.S.
Real annual growth rate (2008):	3.1%
GDP per capita (2008):	$9,686 U.S.
Goods exported:	Fish products, reexport of petroleum products
Goods imported:	Food, machinery, petroleum products, transportation equipment
Trading partners:	France, Germany, Saudi Arabia, Spain, United Kingdom

A shortage of suitable farmland limits crop production. Cinnamon grows wild on much of Mahé, and coconut palms flourish on many of the islands. Bananas, cinnamon, and coconuts are the nation's chief products.

But the basis of the Seychelles economy is tourism. The country's remote location and beautiful beaches—as well as its exotic plant and animal life, such as double coconuts and giant tortoises—attract vacationers.

The tourism industry and the government are the two main employers in Seychelles. A construction industry developed in the 1970's to meet the demand for hotels and restaurants. Many other people work in the farming and manufacturing industries.

A tropical beach on one of the coral islands displays the striking natural beauty that has made tourism the basis of the Seychelles economy.

Seychelles is an African country that consists of about 115 islands scattered about 1,000 miles (1,600 kilometers) east of the mainland in the Indian Ocean. Many of the islands are uninhabited.

A trader displays his cocos de mer, the double coconuts weighing up to 50 pounds (23 kilograms) that grow only on the Seychelles.

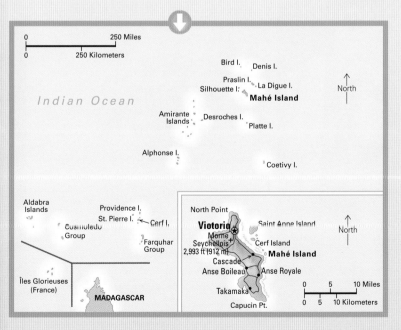

SIERRA LEONE

Sierra Leone sits on the western bulge of Africa, north of the equator. Although it is a small country, Sierra Leone provides a large portion of one of the world's most valuable treasures—diamonds.

The people of Sierra Leone belong to nearly 20 main ethnic groups. About one-third of the people are Mende, who live in the south, and another third are Temne, who live mainly in the west. A small percentage of the people are Creoles, who are descended from freed slaves. English is the official language of Sierra Leone, but most of the people speak local African languages. The Creoles speak *Krio,* which is derived from English and African languages.

Most men in Sierra Leone are farmers, but they grow only enough food for their families. During the dry season—January and February in the south and December through March in the north—the men mine diamonds. Many women in Sierra Leone sell goods in local markets.

Gravel or sandy soil covers more than half the land. Because of the poor soil, the dry season, and old-fashioned farming methods, crop yields are low. The nation's main food crops are cassava, oranges, peanuts, rice, and tomatoes. Cash crops include cacao, which is used to make chocolate; kola nuts, which are used to make soft drinks; palm kernels, which contain a valuable oil; and piassava, a fiber used to make brushes.

But diamonds account for about half of the total value of Sierra Leone's exports. The precious stones lie in gravel deposits along riverbeds and in swamps in the eastern part of the country. About 70 percent of the diamonds are used as gemstones, and the others are used in industry.

Sierra Leone is a former British colony. Beginning in the 1500's, Europeans enslaved many Africans from the region and shipped them to America. In 1787, Granville Sharp, an Englishman opposed to slavery, helped about 400 freed black slaves settle on land where Freetown, Sierra Leone's capital and largest town, now stands.

FACTS

Official name:	Republic of Sierra Leone
Capital:	Freetown
Terrain:	Coastal belt of mangrove swamps, wooded hill country, upland plateau, mountains in east
Area:	27,699 mi² (71,740 km²)
Climate:	Tropical; hot, humid; summer rainy season (May to December); winter dry season (December to April)
Main rivers:	Great and Little Scarcies, Jong, Moa, Sewa
Highest elevation:	Loma Mansa, 6,390 ft (1,948 m)
Lowest elevation:	Atlantic Ocean, sea level
Form of government:	Constitutional democracy
Head of state:	President
Head of government:	President
Administrative areas:	3 provinces, 1 area
Legislature:	House of Representatives with 124 members serving five-year terms
Court system:	Supreme Court
Armed forces:	10,500 troops
National holiday:	Independence Day - April 27 (1961)
Estimated 2010 population:	6,276,000
Population density:	227 persons per mi² (87 per km²)
Population distribution:	62% rural, 38% urban
Life expectancy in years:	Male, 43; female, 46
Doctors per 1,000 people:	Less than 0.05
Birth rate per 1,000:	46
Death rate per 1,000:	22
Infant mortality:	155 deaths per 1,000 live births
Age structure:	0-14: 43%; 15-64: 54%; 65 and over: 3%
Internet users per 100 people:	0.3
Internet code:	.sl
Languages spoken:	English (official), Mende, Temne, Krio
Religions:	Muslim 60%, indigenous beliefs 30%, Christian 10%
Currency:	Leone
Gross domestic product (GDP) in 2008:	$1.96 billion U.S.
Real annual growth rate (2008):	6.0%
GDP per capita (2008):	$331 U.S.
Goods exported:	Mostly diamonds, also cocoa, coffee
Goods imported:	Food, machinery, motor vehicles, petroleum products
Trading partners:	Belgium, China, Côte d'Ivoire, Netherlands, United States

In 1807, the United Kingdom outlawed the slave trade, and the following year it made the Sierra Leone Peninsula a British colony. The British then freed slaves from the slave ships of many nations and settled them in the colony. British influence gradually spread inland, and in 1896 the United Kingdom established a protectorate over the whole region.

Sierra Leone moved gradually toward self-government. In 1961, it became a completely independent nation with a government headed by a prime minister. In 1971, Sierra Leone became a republic with a president.

In 1991, Sierra Leone adopted a constitution that allowed multiparty elections. But in 1992, soldiers led by Captain Valentine Strasser took control of the government and canceled the elections. Corporal Foday Sankoh had also led an uprising against the government. During the 1990's, the government changed hands several times, falling under both civilian and military rule. Ahmed Tejan Kabbah, a civilian president, ended up in office. Resistance from Sankoh's forces, known as the Revolutionary United Front (RUF), continued.

The former residence of a British official nestles amid palm trees and other tropical vegetation. Sierra Leone was a British possession until 1961.

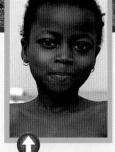

Sierra Leone is a small country in western Africa. Swamps cover most of its coastal region. Inland, a plain slopes upward to the northeastern plateaus, and mountains that cover about half the country.

Sierra Leone children are not required by law to go to school. About 40 percent attend elementary school.

In 1999, Kabbah and Sankoh signed a peace agreement. Under the accord, Sankoh was given a position in the Sierra Leone government. The agreement also called for the rebels to disarm, but they refused to do so.

In early May 2000, the rebels seized about 500 United Nations (UN) peacekeepers and held them hostage. Sankoh went into hiding and was captured after a violent protest outside his house. In 2001, the UN began disarming the rebels. In 2002, the war was declared over, and Kabbah was reelected president. Ernest Bai Koroma was elected president in 2007.

In early 2009, a UN-backed international court found three senior RUF leaders guilty of war crimes and crimes against humanity.

SINGAPORE

Singapore is a small island country that lies near the southern tip of the Malay Peninsula. The nation consists of a large island and more than 50 smaller islands.

Rain forests cover most of the central part of the main island, and mangrove swamps lie along the northern coast. Monkeys, snakes, and some other reptiles still live there, but many other animals, such as tigers and leopards, are no longer found in Singapore due to urban development.

A Western creation

Until the early 1800's, Singapore was a land of jungles and swamps with only small fishing villages on the main island. Sir Stamford Raffles, an agent for a British trading organization called the East India Company, realized that the island's ideal location for trade could benefit his country. In 1819, he gained possession of Singapore harbor for the United Kingdom through an agreement with the sultan of Johor and a local chief of Singapore.

In 1824, all of Singapore came under British control. As the port prospered, Singapore's population grew rapidly. Large numbers of Chinese settled in Singapore, and many became merchants.

In the 1920's and 1930's, the British built air bases and a naval base on Singapore Island. Nevertheless, Japanese forces captured Singapore during World War II (1939-1945). They occupied the island until 1945, when the British regained control.

Singapore became part of Malaysia in 1963, and it gained independence in 1965. Although political freedom is limited, the country's economy and population have grown steadily. With its modern high-rises and busy streets, this crowded, bustling center of trade has come a long way from its jungle origins.

Politics and government

Singapore is a republic. A Parliament consisting of a single house makes the country's laws. A prime minister and a cabinet carry out the operations of the government. A president serves as head of state but has little actual power.

FACTS

Official name:	Republic of Singapore
Capital:	Singapore
Terrain:	Lowland; gently undulating central plateau contains water-catchment area and nature preserve
Area:	270 mi² (699 km²)
Climate:	Tropical; hot, humid, rainy; no pronounced rainy or dry seasons; thunderstorms occur often, especially in April
Main rivers:	Jurong, Kallang, Singapore
Highest elevation:	Timah Hill, 581 ft (177 m)
Lowest elevation:	Singapore Strait, sea level
Form of government:	Parliamentary republic
Head of state:	President
Head of government:	Prime minister
Administrative areas:	None
Legislature:	Parliament with 84 members serving five-year terms
Court system:	Supreme Court, Court of Appeals
Armed forces:	72,500 troops
National holiday:	National Day - August 9 (1965)
Estimated 2010 population:	4,701,000
Population density:	17,411 persons per mi² (6,725 per km²)
Population distribution:	100% urban
Life expectancy in years:	Male, 79; female, 85
Doctors per 1,000 people:	1.5
Birth rate per 1,000:	10
Death rate per 1,000:	5
Infant mortality:	2 deaths per 1,000 live births
Age structure:	0-14: 18%; 15-64: 73%; 65 and over: 9%
Internet users per 100 people:	70
Internet code:	.sg
Languages spoken:	Mandarin Chinese, English, Malay, Tamil (all official)
Religions:	Buddhist and Chinese traditional religions 51%, Muslim 14.9%, Christian 14.6%, Hindu 4%, other 15.5%
Currency:	Singapore dollar
Gross domestic product (GDP) in 2008:	$181.94 billion U.S.
Real annual growth rate (2008):	1.2%
GDP per capita (2008):	$40,730 U.S.
Goods exported:	Electrical and electronic equipment, machinery, petroleum products, telecommunications equipment
Goods imported:	Chemicals, electrical goods, food, machinery and equipment, petroleum
Trading partners:	China, Japan, Malaysia, South Korea, Taiwan, Thailand, United States

The head of the political party with the most seats in Parliament serves as the prime minister of the country. The People's Action Party (PAP), which is Singapore's largest political party, held all the seats in Parliament from 1968 until 1981. Since then, the PAP has held almost all the seats.

Economic development

Singapore has a highly developed economy. Import and export activities provide jobs for many people in the port area. Singapore has few natural resources, but it is a major manufacturing center. Singapore's factories produce such goods as chemicals, electronic equipment, machinery, rubber and plastic products, telecommunications equipment, televisions, and transportation equipment.

Singapore has also become a major financial center with many banks, insurance and finance companies, and a stock exchange. Few people are unemployed in Singapore. The nation's annual income *per capita* (per person) is one of the highest in Asia.

Singapore, the nation's capital city, lies on the southern coast of Singapore Island. It has one of the busiest ports in the world. Ships sailing to Australia, China, and Japan dock there to load and unload cargo. Singapore is a *free port*—that is, a port where goods can be unloaded, stored, and reshipped without payment of import duties.

The skyline of Singapore's financial district towers over the Singapore River. The Elgin Bridge stands in the foreground. The bridge, completed in 1929, was the first to span the river.

Singapore is a small island country in Southeast Asia. It lies at the southern tip of the Malay Peninsula. Singapore consists of a large island, called Singapore, and more than 50 smaller islands. About half are uninhabited.

Keppel Harbor is one of Southeast Asia's busiest ports.

PEOPLE

People have lived on what is now Singapore island since prehistoric times. By 1400, the city had become known as *Singapura,* which means *lion city* in the Sanskrit language. However, the origin of the name is uncertain.

During the 1300's, the harbor in Singapore served as a trading center. Invaders from Java (now part of Indonesia) destroyed much of the island in 1377. Singapore then became a base for pirates and fishing fleets. Until the early 1800's, most of the area was covered by jungles and swamps.

Modern-day Singapore, which is still a trading center, has a population of more than 4.7 million. It ranks among the world's most densely populated countries, with an average of 17,411 people per square mile (6,725 per square kilometer). Among Singapore's people, more than 75 percent are Chinese, about 15 percent are Malays, and the remaining 10 percent are mainly East Indians.

Contrast and unity

Almost all of the country's people live in the capital city of Singapore, which is built around the harbor and offers a wide variety of architecture and housing. In the crowded downtown section, modern skyscrapers tower over traditional Chinese shops.

Near the docks, Singapore's older residential sections have many brick and stucco shops with living quarters on the second floor. Some Malays live in bamboo houses thatched with palm leaves. In the wealthier sections of the city, people live in modern single-family homes with well-kept flower gardens, shrubs, and trees. Since the mid-1900's, the government has built communities called *new towns,* which consist of many large apartment complexes, on the main island.

Singapore's ethnic groups create a variety of cultures within the country. For example, Singapore has four official languages—Mandarin Chinese, English, Malay, and Tamil. Most of the Chinese practice Buddhism or Taoism, while most Malays are Muslims. Hinduism ranks as the main religion among the Indians of Singapore, and Christianity among the Europeans.

Most urban Singaporeans wear clothing similar to that worn in Europe and North America. However, some Indi-

A Chinese calligrapher who earns his living by the ancient art of beautiful handwriting, waits for the ink to dry on his latest piece of work. Many traditional Chinese shops sit in the shadows of Singapore's modern skyscrapers.

Kampong housing was once common in Singapore. However, few of these traditional buildings remain.

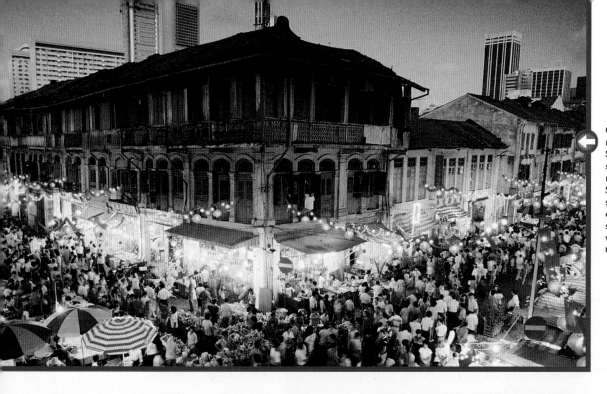

Crowded street markets in Singapore attract shoppers from many countries. Street vendors sell such wares as batik cloth, wood and stone carvings, silver, and antique Chinese porcelain.

ans and Malays prefer traditional dress. Restaurants offer an assortment of Chinese, Indian, and Malay dishes. The city's art, music, and theater reflect the cultures of Singapore's various ethnic groups, with Chinese opera, Indian dancing, and Malay dramas.

Despite ethnic differences, relations among the major groups are fairly peaceful. Singapore has no official religion, and the government has worked to develop a sense of national identity that goes beyond any specific ethnic group.

A prosperous society

The government of Singapore exercises strong control over the nation's economy, and the country and the people have prospered. The government determines how much vacation time and sick leave employers must provide for their workers. In addition, the government operates an employment agency to help people find jobs, and it provides pensions for retired workers.

The English language is used in all Singapore's schools. In addition to English, the government requires children to learn one of the other official languages, usually the native language of the child's family. Singapore's children are given six years of primary education and four or five years of secondary school. About half of the people in Singapore are able to read two or more languages. Schools of higher education include the National University of Singapore and Nanyang Technological Institute.

The luxurious Raffles Hotel dates from the late 1880's. It is named after Sir Stamford Raffles, who gained possession of Singapore for the United Kingdom in 1819.

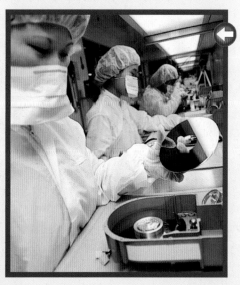

Many Singaporeans are involved in the manufacturing of electronic equipment, which has brought Singapore great wealth. Singapore has few natural resources, so it must import most of the raw materials needed for manufacturing.

SLOVAKIA

Situated in the heart of central Europe is the small, land-locked country of Slovakia. It is bordered to the north-west by the Czech Republic, to the north by Poland, to the east by Ukraine, to the south by Hungary, and to the west by Austria.

From 1918 to 1992, Slovakia was part of Czechoslovakia. Slovakia became an independent nation on Jan. 1, 1993, when Czechoslovakia was dissolved.

Slovakia is divided into two major land regions: the Danubian Lowlands and the Western Carpathians. The Danubian Lowlands, in southwestern Slovakia, have fertile farmland, though some areas are too marshy for crops. Most of the remainder of Slovakia consists of the Western Carpathians, a thinly populated region with a rugged landscape of forested mountain ranges.

History

Slavic tribes settled in what is now Slovakia beginning in the A.D. 400's. During the 800's, the area became part of the Greater Moravian Empire. The *Magyars* (Hungarians) conquered the empire in 907 and ruled the Slovaks for the next 1,000 years. Although the Slovaks preserved their own language under Hungarian rule, they lacked a strong sense of national identity until the late 1700's.

In 1867, Austria and Hungary formed a monarchy called Austria-Hungary. After World War I (1914-1918), Austria-Hungary was split up into several nations, and the Slovaks united with the Czechs to form the republic of Czechoslovakia. Because most Slovaks were peasant farmers with little political experience, the Czechs took control of the new nation's economy and government. But the Slovaks grew to resent the Czechs' power. In 1948, Communists took control of the government.

In 1989, protests by Czechs and Slovaks led to the resignation of the Communist government. Non-Communists came to power after elections the following year. In 1992, when talks over Slovakia's demands for sovereignty within the Czechoslovak federation failed, leaders of both republics moved forward to split the country into two separate nations. On Jan. 1, 1993, Slovakia was formally established as an independent country. Slovakia joined both the European Union and the North Atlantic Treaty Organization in 2004.

FACTS

Official name:	Slovenska Republika (Slovak Republic)
Capital:	Bratislava
Terrain:	Rugged mountains in the central and northern part; lowlands in the south
Area:	18,933 mi² (49,035 km²)
Climate:	Temperate; cool summers; cold, cloudy, humid winters
Main river:	Danube
Highest elevation:	Gerlachovsky Stit, 8,711 ft (2,655 m)
Lowest elevation:	308 ft (94 m), near the Bodrog River on the Hungarian border
Form of government:	Parliamentary democracy
Head of state:	President
Head of government:	Prime minister
Administrative areas:	8 kraje (regions)
Legislature:	Narodna Rada Slovenskej Republiky (National Council of the Slovak Republic) with 150 members serving four-year terms
Court system:	Supreme Court
Armed forces:	17,400 troops
National holiday:	Constitution Day - September 1 (1992)
Estimated 2010 population:	5,406,000
Population density:	286 persons per mi² (110 per km²)
Population distribution:	56% urban, 44% rural
Life expectancy in years:	Male, 71; female, 79
Doctors per 1,000 people:	3.1
Birth rate per 1,000:	10
Death rate per 1,000:	10
Infant mortality:	7 deaths per 1,000 live births
Age structure:	0-14: 16%; 15-64: 72%; 65 and over: 12%
Internet users per 100 people:	51
Internet code:	.sk
Languages spoken:	Slovak (official), Hungarian, Roma, Ukrainian
Religions:	Roman Catholic 68.9%, Protestant 10.8%, Greek Orthodox 4.1%, other 16.2%
Currency:	Euro
Gross domestic product (GDP) in 2008:	$96.99 billion U.S.
Real annual growth rate (2008):	6.4%
GDP per capita (2008):	$17,977 U.S.
Goods exported:	Chemicals, electrical equipment, iron and steel, machinery, motor vehicles
Goods imported:	Electrical equipment, machinery, petroleum and petroleum products, transportation equipment
Trading partners:	Austria, Czech Republic, Germany, Poland, Russia

A quiet village, set amid rolling hills and fertile farmland, reflects Slovakia's mostly rural landscape. The country's best farmland lies in the southeast region, in the Danube River valleys.

Two traders meet in a market in Bratislava. The city is a major transportation center. Its factories make ceramics, chemicals, machinery, and petroleum products.

Economy

Slovakia was once largely an agricultural region. Since the late 1940's, it has become industrialized, and about a third of the people work in industry. Service industries have developed rapidly since the fall of Communism and now employ about half of the country's workers.

About a tenth of the Slovak people work in agriculture. Farmers raise barley, corn, potatoes, sugar beets, and wheat. The country also has deposits of copper, iron, lead, manganese, and zinc.

People

Slovaks form the majority of Slovakia's population, and Slovak is the country's official language. People of Hungarian descent make up the second largest ethnic group. Roma (sometimes called Gypsies) are another significant minority group.

Slovakia has a higher standard of living than many other formerly Communist countries in Europe. Most Slovak families own automobiles, refrigerators, televisions, and washing machines.

Typical foods include *bryndzove halusky* (noodles with sheep's cheese) and *goulash* (paprika-flavored stew). Wine and a plum brandy called *slivovice* are common alcoholic beverages.

Slovakia, which became an independent nation in 1993, lies in the heart of central Europe. Bratislava, a major port on the Danube River, is Slovakia's capital and largest city.

THE DANUBE RIVER

The Danube—the second longest river in Europe, after the Volga—flows 1,770 miles (2,850 kilometers) from its source in the Black Forest of Germany. In its journey to the sea, the mighty Danube travels through Germany, Austria, and eastern Europe. The river ends in Romania, where it empties into the Black Sea. It has a number of major ports.

A historic waterway

Throughout history, the Danube has been one of Europe's great waterways and the site of many battles. For the early Romans, the mighty river served as a boundary between their empire and the barbarians. The fortified towns and settlements they erected along its banks became such cities as Vienna, Austria; Budapest, Hungary; and Belgrade, Serbia.

The Holy Roman Empire and, later, the Ottoman Empire used the Danube as a route for conquest as well as a means of defense. Along its banks, history has witnessed the rise and fall of the Byzantines, Goths, and Huns, as well as Napoleonic imperialism and Stalinist Communism.

The river begins where two small rivers, the Breg and the Brigach, meet in the Black Forest of southwestern Germany. From there, the Danube flows northeast to Regensburg before turning southeast to enter Austria at Passau. There, the waters of an important tributary, the Inn, increase the river's size.

The Danube continues eastward to form part of the border between Slovakia and Hungary, and then turns south toward central Hungary, flowing through the capital city of Budapest. The river passes Belgrade before entering Romania through the spectacular Iron Gate gorge, where, after swelling from numerous tributaries, it squeezes through Europe's deepest gorge. After passing through the Iron Gate, the mighty waters tumble out onto the Danubian plain in a series of five spectacular waterfalls.

The New Bridge crosses the Danube at Bratislava. The bridge, completed in 1972, has four upper lanes for automobiles and lanes for bicycles and pedestrians beneath the roadway. The saucer-shaped structure atop the piers contains a restaurant.

Bratislava Castle overlooks the Danube River. A castle has stood on this spot as far back as the 800's. Bratislava, a major port on the Danube, is Slovakia's capital and largest city.

Journey to the Black Sea

In Romania, the river eventually turns northeast before shifting to the east and splitting into the many branches of the Danube Delta. The northernmost part of the delta forms the boundary between Ukraine—the former Soviet republic—and Romania.

The Danube has no single mouth. The many channels of the Danube Delta carry so much silt that the delta is moving out into the Black Sea at the rate of about 100 feet (30 meters) a year.

The Danube is used by all the countries along its course for the transport of goods, and many busy ports have developed along its banks. Ships can sail on the Danube all the way from the North Sea to the Black Sea by using the 106-mile-long (171-kilometer-long) Main-Danube Canal.

In the 1990's, Slovakia's construction of a dam at Gabcikovo, southeast of Bratislava, drew significant criticism. Environmental groups claimed that the dam would destroy wildlife, flood valuable land, and damage one of Europe's largest underground water supplies. Countries situated downstream of the dam argued that it would interfere with the Danube's flow and disrupt river transport.

The Danube River is the second longest river in Europe. It begins at the merger of two small rivers in the Black Forest in Germany and empties into the Black Sea.

SLOVENIA

Slovenia is a mountainous and densely forested land in central Europe at the northernmost tip of the Adriatic Sea. The republic is bordered by Italy, Austria, Hungary, and Croatia. Slovenia, formerly one of the six federal republics of Yugoslavia, withdrew from the federation in 1991 and became an independent nation.

Almost all of Slovenia's people are Slovenes, a Slavic people who speak Slovenian. Most are Roman Catholic. About half of the people live in rural areas, and the remainder live in cities and towns.

The ancestors of present-day Slovenes settled in what is now Slovenia during the late A.D. 500's. By the 800's, Germanic tribes had invaded and gained control of the region, and the majority of the Slovenes were converted to Christianity. Under German rule, the Slovenes became part of the Holy Roman Empire. Austrians conquered the area in the late 1200's. Present-day Slovenia became the province of Carniola under Austrian rule, which lasted until 1918. As a result, the Slovenes considered themselves to be part of Western European culture.

After World War I (1914-1918), Slovenia became part of a new nation called the Kingdom of the Serbs, Croats, and Slovenes, which was renamed Yugoslavia in 1929. During World War II (1939-1945), Germany and Italy conquered Slovenia and divided the republic with Hungary. After the war, Slovenia became one of the six republics of the Socialist Federal Republic of Yugoslavia, which was controlled by the Communist Party.

Slovenia was the most developed region in Yugoslavia, with the highest income *per capita* (per person). Eventually, the people of Slovenia began to resent that a relatively large amount of the central government's budget went to help develop the poorer Yugoslav republics.

In the late 1980's, the secessionist movement in Slovenia was also fueled by Slovene concerns over Serbian nationalism and the Serbs' treatment of ethnic Albanians in the Serbian province of Kosovo. In 1989, the Slovene Assembly passed an amendment to the republic's constitution, establishing the right of the Slovene people to secession and union. The Serbs responded by conducting an economic boycott against Slovenia.

FACTS

Official name:	Republika Slovenija (Republic of Slovenia)
Capital:	Ljubljana
Terrain:	A short coastal strip on the Adriatic, an Alpine mountain region adjacent to Italy and Austria, mixed mountains and valleys with numerous rivers to the east
Area:	7,827 mi² (20,273 km²)
Climate:	Mediterranean climate on the coast, continental climate with mild to hot summers and cold winters in the plateaus and valleys to the east
Main rivers:	Dravinja, Savinja
Highest elevation:	Mount Triglav, 9,393 ft (2,863 m)
Lowest elevation:	Adriatic Sea, sea level
Form of government:	Parliamentary democratic republic
Head of state:	President
Head of government:	Prime minister
Administrative areas:	182 obcine (municipalities), 11 obcine mestne (urban municipalities)
Legislature:	Parliament consists of the Drzavni Svet (National Council) with 40 members serving five-year terms and the Drzavni Zbor (National Assembly) with 90 members serving four-year terms
Court system:	Supreme Court, Constitutional Court
Armed forces:	7,200 troops
National holiday:	Independence Day (Statehood Day) - June 25 (1991)
Estimated 2010 population:	2,008,000
Population density:	257 persons per mi² (99 per km²)
Population distribution:	52% rural, 48% urban
Life expectancy in years:	Male, 74; female, 81
Doctors per 1,000 people:	2.4
Birth rate per 1,000:	10
Death rate per 1,000:	9
Infant mortality:	3 deaths per 1,000 live births
Age structure:	0-14: 14%; 15-64: 70%; 65 and over: 16%
Internet users per 100 people:	50
Internet code:	.si
Languages spoken:	Slovenian, Serbo-Croatian
Religions:	Catholic 57.8%, Muslim 2.4%, Orthodox 2.3%, other 37.5%
Currency:	Euro
Gross domestic product (GDP) in 2008:	$54.64 billion U.S.
Real annual growth rate (2008):	4.3%
GDP per capita (2008):	$27,306 U.S.
Goods exported:	Chemicals, furniture, iron and steel, machinery, transportation equipment
Goods imported:	Chemicals, food, machinery, petroleum products, transportation equipment
Trading partners:	Austria, Croatia, France, Germany, Italy

In 1990, the Communists ended their monopoly on political power in Yugoslavia. In multiparty elections held in Slovenia that year, non-Communist parties won a majority of parliamentary seats.

In June 1991, along with its neighboring republic, Croatia, Slovenia declared its independence. The Yugoslav national government strongly opposed the move, and fighting broke out between Slovenian and Croatian troops and the Yugoslav military. As part of a cease-fire plan established in early July, Slovenia agreed to postpone its declaration of independence for three months. When the postponement expired in October 1991, Slovenia went ahead with its plans for independence.

In the late 1990's and early 2000's, Slovenia worked toward joining the European Union (EU) and the North Atlantic Treaty Organization (NATO). Slovenia became a member of both the EU and NATO in 2004.

When Slovenia withdrew from the Yugoslav federation, it had a stronger economy and higher standard of living than any other Yugoslav republic. Today, most of the country's workers are employed in service industries, and many others hold jobs in manufacturing and agriculture. Automobiles, chemicals, metal goods, and textiles are the chief manufactured products. Major crops include corn, potatoes, and wheat.

The majestic Julian Alps form a spectacular backdrop to a farm in Slovenia. Narrow valleys and deep gorges cut through these towering summits.

Slovenia is bordered by Italy to the west, Austria and Hungary to the north, and Croatia to the south. Ljubljana is its capital and largest city.

SOLOMON ISLANDS

FACTS

The Solomon Islands is an island country about 1,000 miles (1,610 kilometers) northeast of Australia, in the South Pacific Ocean. The country's largest islands are part of an island chain that is also called the Solomon Islands. But not all the islands in the chain belong to the country. Some islands in the northern part of the chain are part of Papua New Guinea. The Solomon Islands are considered part of Melanesia.

The country covers about 230,000 square miles (600,000 square kilometers) of the Pacific, but it has a land area of only 11,157 square miles (28,896 square kilometers). The country's main islands, which were formed by volcanoes, are rugged, mountainous, and covered with tropical plants. Each has a central spine of mountains. Some mountains rise more than 4,000 feet (1,200 meters) high. The land drops sharply to the sea on one side of the islands and slopes gently down to a narrow coastal strip on the other. Some of the outlying islands are atolls.

Government and economy

The Solomon Islands is a parliamentary democracy and a member of the Commonwealth of Nations. Parliament elects one of its own members as prime minister, the head of government. A cabinet, appointed by the prime minister, helps run the government. A 50-member Parliament, elected by the people to four-year terms, makes the country's laws. Honiara, on Guadalcanal, is the capital and largest community of the Solomons.

Fish, timber, palm oil, cocoa, and copra are the country's main products. Japan buys much of the fish and timber the country exports. The country has good ferry and shipping services, but it has few roads. Air routes connect the Solomon Islands with Australia and neighboring islands. The government publishes a weekly newspaper and broadcasts radio programs.

People and history

Most Solomon Islanders are Melanesians, and about 80 percent live in rural villages. Many people build their houses on stilts to keep the dwellings cool. Although English is the official language of the country,

Official name:	Solomon Islands
Capital:	Honiara
Terrain:	Mostly rugged mountains with some low coral atolls
Area:	11,157 mi² (28,896 km²)
Climate:	Tropical monsoon; few extremes of temperature and weather
Main rivers:	N/A
Highest elevation:	Mount Makarakomburu, 8,028 ft (2,447 m)
Lowest elevation:	Pacific Ocean, sea level
Form of government:	Parliamentary democracy
Head of state:	British monarch, represented by governor general
Head of government:	Prime minister
Administrative areas:	9 provinces, 1 capital territory
Legislature:	National Parliament with 50 members serving four-year terms
Court system:	Court of Appeal
Armed forces:	None
National holiday:	Independence Day - July 7 (1978)
Estimated 2010 population:	533,000
Population density:	48 persons per mi² (18 per km²)
Population distribution:	82% rural, 18% urban
Life expectancy in years:	Male, 67; female, 70
Doctors per 1,000 people:	0.1
Birth rate per 1,000:	31
Death rate per 1,000:	6
Infant mortality:	48 deaths per 1,000 live births
Age structure:	0-14: 39%; 15-64: 58%; 65 and over: 3%
Internet users per 100 people:	1.8
Internet code:	.sb
Languages spoken:	Solomons Pidgin, English (official), more than 100 indigenous languages
Religions:	Christian 95%, other 5%
Currency:	Solomon Islands dollar
Gross domestic product (GDP) in 2008:	$473 million U.S.
Real annual growth rate (2008):	7.3%
GDP per capita (2008):	$927 U.S.
Goods exported:	Cocoa, copra, fish, palm oil, timber
Goods imported:	Food, machinery, motor vehicles, petroleum products
Trading partners:	Australia, China, Japan, Singapore, South Korea

more than 100 languages are spoken among the Melanesians. The islanders also speak *Pidgin English,* a dialect based on English that is often used among people with no common language. The government uses both English and Pidgin English in its broadcasts and publications.

According to archaeologists, the Solomon Islands were first settled about 6,000 years ago by people from New Guinea. In 1568, a Spanish explorer named Álvaro de Mendaña became the first European to reach the islands. Then, from 1870 to 1911, Europeans recruited nearly 30,000 Solomon Islanders to work on plantations in Fiji and in Queensland, Australia.

The United Kingdom took control of the Solomons in 1893. Guadalcanal and other islands were the scene of fierce fighting between Allied and Japanese forces in 1942 and 1943, during World War II. Many people from Malaita Island moved to Guadalcanal Island during and after the war.

The Solomon Islands gained independence from the United Kingdom on July 7, 1978. Through the years, resentment between the Malaitans and native Guadalcanal islanders grew, and fighting broke out in the early 2000's. In 2003, Australia led a multinational peace-keeping force to the islands. The force restored order by mid-2004. In 2009, the United Nations established a Truth and Reconciliation Commission to conduct hearings into the violence that occurred between the Malaitans and native Guadalcanal islanders from 1997 to 2003.

Rusting artillery and aircraft on Guadalcanal are a grim reminder of the fighting that took place there during World War II.

Melanesians make up most of the population of the Solomon Islands.

An outrigger canoe skips over the waves in a lagoon in the southern Solomon Islands.

The Solomon Islands is a country in the South Pacific Ocean. The nation's largest islands are part of the Solomon Islands chain, but not all islands in the chain are part of the country.

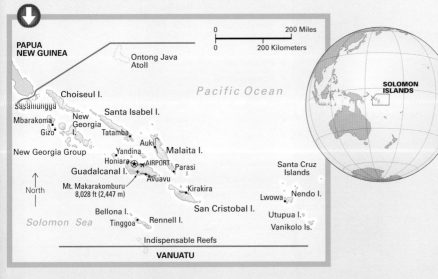

PAPUA NEW GUINEA

Ontong Java Atoll

0   200 Miles
0   200 Kilometers

Pacific Ocean

Choiseul I.

Sasamungga

Santa Isabel I.

Mbarakoma
New Georgia I.
Gizo   Tatamba

New Georgia Group

Yandina
Auki
Malaita I.

Honiara ⊕ ✈ AIRPORT
Guadalcanal I.   Parasi
Avuavu

North

Mt. Makarakomburu
8,028 ft (2,447 m)

Kirakira

Santa Cruz Islands

Lwowa   Nendo I.

Bellona I.

San Cristobal I.

Solomon Sea   Tinggoa   Rennell I.

Utupua I.

Vanikolo Is.

Indispensable Reefs

VANUATU

SOLOMON ISLANDS

SOMALIA

Somalia is the country on the "horn of Africa"—a piece of land that juts eastward into the Indian Ocean. Dry, grassy plains called *savannas* cover almost all of Somalia, though a mountain ridge rises behind a narrow coastal plain in the north.

Most of the grasslands are suitable only for grazing livestock, and the Somali economy has long been based on the herding of camels, cattle, goats, and sheep. In the north, some Somalis fish for a living. In the south, where rivers provide water for irrigation, farmers grow bananas, corn, sorghum, and sugar cane. The country has a few light industries based mainly on its agricultural activities, like refining sugar and milling cotton.

Most Somalis are nomadic herders who wander across the hot, dry country in search of pasture, living in small, collapsible huts. Like their ancestors, they rely heavily on camels for transportation, food, and clothing.

The nomads belong to four clans called, collectively, the Samaal. Two other clans, called the Sab, live in the south and make their living as farmers. For years, most Somalis were loyal only to their clan. But today, the Samaal and the Sab think of themselves as one people. Almost all Somalis speak Somali and follow Islam.

In the mid-1800's, the United Kingdom took over northern Somalia, which became British Somaliland. Italy gained control of most of the Indian Ocean coast, then advanced inland and established the colony of Italian Somaliland. By the early 1900's, Somali nationalists led by Sayyid Muhammad Abdille Hassan were fighting British, Italian, and Ethiopian forces.

During World War II (1939-1945), the United Kingdom took control of Italian Somaliland. In 1950, the United Nations (UN) ruled that the territory should be granted independence in 10 years. In 1960, both the United Kingdom and Italy granted their territories freedom. On July 1, the new independent state of Somalia was formed. In 1969, army officers led by Major General Mohamed Siad Barre seized control of Somalia.

FACTS

Official name:	Somalia
Capital:	Mogadishu
Terrain:	Mostly flat plateau rising to hills in north
Area:	246,201 mi² (637,657 km²)
Climate:	Principally desert; northeast monsoon, moderate temperatures in north and very hot in south (December to February); southwest monsoon, torrid in the north and hot in the south, irregular rainfall, hot and humid periods between monsoons (May to October)
Main rivers:	Jubba, Shabeelle
Highest elevation:	Mount Surud Ad, 7,900 ft (2,408 m)
Lowest elevation:	Indian Ocean, sea level
Form of government:	Transitional
Head of state:	President
Head of government:	Prime minister
Administrative areas:	18 regions
Legislature:	Transitional 550-member national assembly
Court system:	Not functioning
Armed forces:	None
National holiday:	Foundation of the Somali Republic - July 1 (1960)
Estimated 2010 population:	9,484,000
Population density:	39 persons per mi² (15 per km²)
Population distribution:	63% rural, 37% urban
Life expectancy in years:	Male, 47; female, 50
Doctors per 1,000 people:	Less than 0.05
Birth rate per 1,000:	44
Death rate per 1,000:	17
Infant mortality:	109 deaths per 1,000 live births
Age structure:	0-14: 45%; 15-64: 52%; 65 and over: 3%
Internet users per 100 people:	1.1
Internet code:	.so
Languages spoken:	Somali (official), Arabic, Italian, English
Religion:	Sunni Muslim
Currency:	Somali shilling
Gross domestic product (GDP) in 2008:	$2.60 billion U.S.
Real annual growth rate (2008):	2.6%
GDP per capita (2008):	$289 U.S.
Goods exported:	Charcoal, fish, gold, hides, livestock
Goods imported:	Construction materials, food, petroleum products
Trading partners:	Djibouti, Oman, United Arab Emirates, Yemen

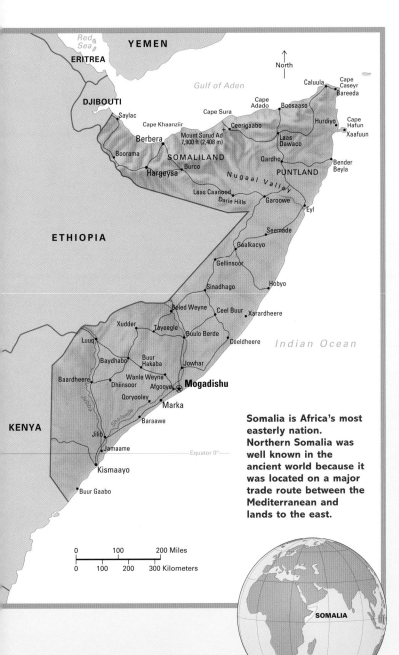

Somalia is Africa's most easterly nation. Northern Somalia was well known in the ancient world because it was located on a major trade route between the Mediterranean and lands to the east.

During the 1970's, famine brought on by drought, along with a war with Ethiopia, caused the deaths of tens of thousands of people. That war ended in 1988, but civil war erupted in northern Somalia between rebel and government forces. Hostility between Somali clans and economic hardship fueled the war. The country was left with no central government. The northern part of Somalia declared its independence as the Somaliland Republic in 1991. Other countries, however, do not recognize its independence.

The fighting and drought again led to widespread starvation in the 1990's. Other countries and international relief groups sent food to Somalia. However, with no central government to protect the food, armed criminals stole much of it. Nearly 300,000 Somalis starved to death. From 1993 to 1995, United Nations peacekeeping troops helped relief groups provide food to needy people.

In 2004, Somali leaders and politicians agreed to establish a new transitional government. Ethiopia sent troops in 2006 to help the transitional government establish itself in Mogadishu. The following year, the African Union (AU) began sending peacekeeping forces to Somalia. However, fighting continued, and in 2009, after an increase in rebel violence, Somalia's president declared a state of emergency. Famine in mid-2011 again claimed thousands of lives.

In the early 2000's, pirates began to establish bases in Somalia. With no effective government to stop them, they wreaked havoc in the Indian Ocean and the Gulf of Aden, where they stopped shipping vessels and private boats and held them for ransom.

A Somali woman walks past a church destroyed by fighting in Mogadishu, the country's capital. Somalia has been plagued by lawlessness and violence throughout the late 1900's and early 2000's.

The traditional clothing of Somalis consists of a piece of brightly colored cloth draped over the head and body. Some urban Somalis wear Western clothes.

SOUTH AFRICA

Like many other African countries, South Africa is a land of dramatic and varied landscapes, ranging from high, sweeping plateaus to towering mountains and deep valleys. Picturesque beaches line its long, fertile coast. In addition, much of South Africa has a delightfully mild and sunny climate.

Unlike most other African countries, however, South Africa is a rich and developed nation. Although South Africa has only about 5 percent of the continent's area and people, it produces 15 percent of the value of the goods and services produced by all African countries combined.

Despite South Africa's abundant resources and beautiful landscape, it was long troubled by violence and isolated from other countries because of its racial policies. It was the last nation in Africa ruled by a white minority. From the late 1940's until the early 1990's, the white-controlled government enforced a policy of rigid racial segregation called *apartheid*. The policy denied voting rights and other rights to the black majority.

In 1990 and 1991, South Africa repealed most of the main laws on which apartheid was based. In 1993, the country extended voting rights to all races, and democratic elections were held the following year. After the 1994 elections, South Africa's white leaders handed over power to the country's first multiracial government. Nelson Mandela, a civil rights leader who had spent 27 years in prison, became South Africa's first black president.

Cape Town, which is South Africa's legislative capital, sits at the foot of Table Mountain. It was established as a Dutch settlement in 1652 and ranks as South Africa's oldest city. Before the Europeans arrived, the region was inhabited by Khoikhoi and San, or Bushmen. After the Dutch came the British and other Europeans.

Today, nearly 80 percent of South Africa's people are black, and about 10 percent are white. Most whites are *Afrikaners,* mainly of Dutch, German, and French descent, and the others are mainly of British descent. *Coloured* people—those of mixed black, white, and Asian descent—also make up about 10 percent of the people. Asians account for less than 5 percent.

SOUTH AFRICA TODAY

The Republic of South Africa is a changing nation. In its first multiracial elections, in 1994, it moved away from apartheid and ushered in a new era of liberation for its black population.

Ending apartheid

Apartheid laws affected many aspects of life in South Africa, including where a person could live or go to school and what jobs a person could hold.

In 1989, Frederik Willem de Klerk was elected South Africa's state president. De Klerk promised to begin undoing apartheid and to establish a "nation without [white] domination." To that end, he released some jailed leaders of the African National Congress (ANC), the main black opposition group, including Nelson Mandela, who had been in prison for 27 years.

In 1990 and 1991, most of the remaining apartheid laws were repealed. Among those repealed were the Black Communities Act, which prohibited blacks from living in white neighborhoods, and the Population Act, which classified all South Africans according to race. In 1993, Mandela and de Klerk were awarded the Nobel Peace Prize for their efforts in bringing about an end to apartheid and in working together to facilitate South Africa's transition to a democracy in which all people, regardless of their race or ethnic group, would participate.

In 1994, South Africa held its first multiracial elections. For the first time, South African blacks were allowed to vote. Amid fears of civil outbreaks, voter turnout was overwhelming, as citizens lined up for hours before the polls opened. In a landslide victory over de Klerk, Mandela became president. It was the first time that Mandela had ever cast a ballot.

South Africa adopted a temporary constitution in 1993. This document provided for a new government that took office in May 1994, following the country's first all-race elections. This government was to serve until 1999. In 1996, South Africa adopted a new constitution. The constitution, which includes a wide-rang-

FACTS

Official name:	Republic of South Africa
Capitals:	Cape Town (legislative), Pretoria (administrative), Bloemfontein (judicial)
Terrain:	Vast interior plateau rimmed by rugged hills and narrow coastal plain
Area:	470,693 mi² (1,219,090 km²)
Climate:	Mostly semiarid; subtropical along east coast; sunny days, cool nights
Main rivers:	Orange, Vaal, Limpopo
Highest elevation:	Njesuthi, 11,181 ft (3,408 m)
Lowest elevation:	Atlantic Ocean, sea level
Form of government:	Parliamentary republic
Head of state:	President
Head of government:	President
Administrative areas:	9 provinces
Legislature:	Parliament consisting of the National Assembly with between 350 and 400 members serving five-year terms and the National Council of Provinces with 90 members serving five-year terms
Court system:	Constitutional Court, Supreme Court of Appeals
Armed forces:	62,100 troops
National holiday:	Freedom Day - April 27 (1994)
Estimated 2010 population:	49,237,000
Population density:	105 persons per mi² (40 per km²)
Population distribution:	60% urban, 40% rural
Life expectancy in years:	Male, 49; female, 50
Doctors per 1,000 people:	0.8
Birth rate per 1,000:	22
Death rate per 1,000:	17
Infant mortality:	45 deaths per 1,000 live births
Age structure:	0-14: 32%; 15-64: 64%; 65 and over: 4%
Internet users per 100 people:	9
Internet code:	.za
Languages spoken:	11 official languages: Afrikaans, English, Ndebele, Sepedi, Sesotho, Swazi, Tsonga, Tswana, Venda, Xhosa, Zulu
Religions:	Christian 79.7%, Muslim 1.5%, other 18.8%
Currency:	Rand
Gross domestic product (GDP) in 2008:	$277.19 billion U.S.
Real annual growth rate (2008):	2.8%
GDP per capita (2008):	$5,883 U.S.
Goods exported:	Diamonds, gold, machinery, motor vehicles, platinum, other metals and minerals
Goods imported:	Food, machinery, petroleum and petroleum products, scientific instruments, transportation equipment
Trading partners:	China, Germany, Japan, United Kingdom, United States

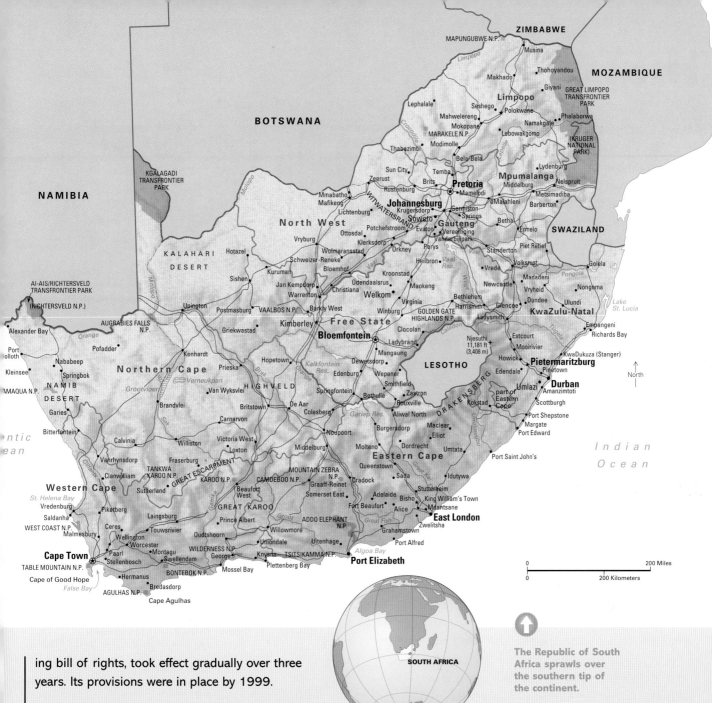

The Republic of South Africa sprawls over the southern tip of the continent.

ing bill of rights, took effect gradually over three years. Its provisions were in place by 1999.

In 1999, Mandela retired as president of South Africa, and Thabo Mbeki was elected president. In 2004, Mbeki was reelected as president. Mbeki stepped down as president in 2008, and Kgalema Motlanthe was elected to replace him. In 2009, Jacob Zuma became president of South Africa.

Today's struggle for equal rights

Today, South Africa has several national institutions designed to strengthen the country's democracy. These institutions include the Human Rights Commission; the Commission for the Promotion and Protection of the Rights of Cultural, Religious and Linguistic Communities; the Commission for Gender Equality; and the Electoral Commission.

However, despite the significant progress made, South Africa's great riches are distributed unevenly among the country's population. White people, who are a minority of South Africa's population, own most of the wealth. Black people, people of mixed race, and people of Asian ancestry—who together make up a majority of the population—own very little.

HISTORY

At least 2,000 years ago, human beings lived in South Africa. In the A.D. 200's, peoples who spoke various Bantu languages began to move into the area from the north. They raised cattle, grew grain, made tools and weapons out of iron, and traded among themselves. By about A.D. 500, the Khoikhoi occupied what is now western South Africa, and Bantu-speaking people occupied much of the eastern part of the country.

In 1652, workers for the Dutch East India Company, led by Jan van Riebeeck, set up a supply base at the present site of Cape Town. In 1657, the company began allowing some employees to start their own farms in Cape Colony. These people became known as *Boers* (farmers). More Dutch farmers followed, as well as French and German settlers. By 1700, whites owned most of the good farmland around Cape Town.

In 1814, the Netherlands gave its Cape Colony to the United Kingdom, and the first British settlers arrived in 1820. However, the Boers soon came to resent British rule. Beginning in 1836, thousands of them made a historic journey called the Great Trek. They loaded their belongings into ox-drawn carts and headed inland into areas occupied by Bantu-speaking peoples, including the Zulu kingdom. The Boers settled in what are now Kwa-Zulu Natal, the Free State, and the Transvaal.

The United Kingdom annexed Natal in 1843 and the Transvaal in 1877. But three years later, the Transvaal Boers revolted. In a conflict called the First Anglo-Boer War (1880-1881), the Boer farmers defeated British troops.

In 1886, a fabulous gold field was discovered in the Transvaal where Johannesburg now stands. British miners and other settlers, called *uitlanders* (foreigners) by the Boers, flocked to the area, and tension between the two groups grew.

Then in 1895, the prime minister of the Cape Colony, Cecil Rhodes, sent troops to overthrow the Transvaal. The invasion failed, and British-Boer relations worsened. In 1899, the Orange Free State joined the Transvaal in declaring war on the United Kingdom. The Boers fought bravely under the leadership of Paulus Kruger, but this time the British were victorious. In 1902, the two Boer republics surrendered and became British colonies.

Cetshwayo (1826–1884)

Paulus Kruger (1825–1904)

Jan Christiaan Smuts (1870–1950)

TIMELINE

About 200 B.C.	Region settled.
A.D. 200's	Black farmers enter eastern South Africa from north.
1488	Portuguese sailors sight Cape of Good Hope.
1652	First Dutch settlers under Jan van Riebeeck arrive at site of Cape Town.
1814	Netherlands gives Cape Colony to the United Kingdom.
1836	Great Trek: Boers leave Cape Colony to occupy Natal, Orange Free State, and Transvaal.
1852	Transvaal becomes a Boer republic.
1854	Orange Free State becomes a Boer republic.
1877	The United Kingdom annexes Transvaal.
1879	The United Kingdom defeats Zulu kingdom.
1880-1881	Transvaal Boers defeat British in First Anglo-Boer War.
1899-1902	The United Kingdom defeats Boers in Second Anglo-Boer War. Boer republics become British colonies.
1910	Union of South Africa created.
1912	African National Congress (ANC) founded.
1920	South Africa gains control over South West Africa (Namibia).
1924	National Party and Labour Party win control of government.
1931	South Africa gains independence.
1948	National Party elected, implements apartheid.
1960	Police kill 69 black protesters at Sharpeville.
1961	South Africa becomes a republic and leaves Commonwealth of Nations.
1964	ANC leader Nelson Mandela sentenced to life imprisonment.
1976	Blacks riot in Soweto; about 600 killed.
1984	Blacks protest against new constitution excluding their role in government.
1986	Trade sanctions imposed on South Africa.
1990	Nelson Mandela released.
1991	Most of the remaining apartheid laws repealed.
1993	Mandela and Frederik Willem de Klerk are awarded the Nobel Peace Prize.
1994	First multiracial elections held; Mandela becomes president.
1999	Mandela retires, and Thabo Mbeki is elected president.
2009	Jacob Zuma is elected president.

A battalion of British troops enters the town of Ladysmith during the Second Anglo-Boer War (1899-1902).

Willem de Klerk, head of South Africa's government from 1989 to 1994, and his successor, Nelson Mandela, right, were instrumental in the ending of apartheid.

Some of the blacks in South Africa—especially the Zulu ruled by Cetshwayo—also resisted British rule. But in 1879, the British defeated the Zulu kingdom, and by 1898, all the black Africans had lost their independence.

The four British colonies formed the Union of South Africa, a self-governing country within the British Empire, in 1910. A great Boer general, Jan Christiaan Smuts, became prime minister of South Africa in 1919. Smuts had fought the British, but as prime minister he tried to unite Afrikaners (as the Boers came to be called) and British South Africans.

Some Afrikaners, however, formed the National Party to promote the belief that the Afrikaners had the right to rule South Africa. When the National Party came to power in 1948, it initiated the policy of apartheid, segregated racial groups, and gave the government great police powers.

Blacks had long opposed the government's racial policies, and their opposition grew with apartheid. The African National Congress (ANC) tried to force reform through boycotts, demonstrations, and strikes, but the government crushed each campaign. The Pan-African Congress (PAC) urged blacks to disobey the law requir-

ing them to carry passes wherever they went. In 1960, at Sharpeville, near Johannesburg, the police killed 69 blacks who were not carrying passes. Almost 600 people were killed in riots in Soweto in 1976.

In the 1970's and 1980's, the government repealed some of the apartheid laws, but others were allowed to stand. A new constitution written in 1984 gave Coloureds and Asians their own houses of Parliament, but blacks were still denied the right to vote. The government declared states of emergency in numerous areas in the late 1980's to deal with protests. Then in 1989, the moderate Willem de Klerk became state president. His government repealed most of the remaining apartheid laws in 1990 and 1991.

PEOPLE

The differing cultural backgrounds of South Africa's people have helped create contrasting ways of life. In addition, the inequalities created by apartheid and by earlier oppression of nonwhites have profoundly affected how people live.

Since the end of apartheid, laws no longer segregate South Africa's racial groups, and there has been much progress toward integration. However, nonwhites still face some unofficial discrimination. Some public elementary and high schools and some housing remain segregated in practice. Also, most high-paying jobs are still held by whites, though blacks and whites do work together in many businesses as the segregation lessens.

The black majority

Blacks make up nearly 80 percent of South Africa's total population. The Zulu people form the largest black ethnic group, followed by the Xhosa, the Sotho, and the Tswana.

Because of years of oppression, many of South Africa's blacks are poor. Large numbers of blacks are unemployed, and some lack adequate housing. Many of South Africa's blacks still live in areas with poor farming soil.

Today, however, nearly 50 percent of blacks live in urban areas. Many are living in formerly segregated neighborhoods, and others have moved into formerly all-white neighborhoods. Still others have built makeshift shelters on empty land inside the city limits and on land along major roads leading into the cities.

Whites

Whites make up 10 percent of South Africa's people. About 60 percent are Afrikaners. Afrikaners are mainly of Dutch descent and generally speak Afrikaans. The other 40 percent of the white population is made up of English-speaking people primarily of British descent.

About 90 percent of all white South Africans live in urban areas and have a high standard of living.

Zulus carrying traditional shields recall the spirit of their ancestors, who fought hard to keep their land in the 1800's. The Zulu are the largest black ethnic group in South Africa today. Many live in eastern Kwa-Zulu Natal.

Ndebele women in traditional dress. The Ndebele is one of South Africa's smaller black ethnic groups.

Two South Africans sharing a bench are a reflection of the improving race relations in the county.

NELSON MANDELA

Nelson Mandela became a powerful symbol of the fight against South Africa's apartheid system. Born the son of a tribal chief in Umtata in 1918, Mandela became a lawyer. In 1944, he joined the African National Congress (ANC), a group opposed to apartheid. He first won national prominence as a leader of black protests in the 1950's. In 1956, Mandela was charged with treason, but he was found not guilty in 1961. In 1962, after renewed protests, he was again arrested and later convicted of sabotage and conspiracy. Mandela was then sentenced to life imprisonment. Over the years, many groups in South Africa and around the world considered Mandela a political prisoner and called for his release. In 1989, Mandela met with the state president of South Africa and expressed his willingness to promote peace between the races in South Africa. In 1990, 27 years after his imprisonment, Nelson Mandela was set free. In 1994, in South Africa's first all-race elections, Mandela was elected president in a landslide victory. He retired as president in 1999.

1918	Born in Umtata.
1944	Joins the ANC.
1950's	Leads black protests.
1956	Brought to trial for treason.
1958	Marries Winifred Nomzamo Madikileza.
1961	Found not guilty.
1962	Arrested on conspiracy Sentenced to life in prison.
1989	Meets with state president.
1990	Released from prison; tours Europe and U.S.
1993	Wins Nobel Peace Prize with Frederik Willem de Klerk.
1994	Elected president of South Africa.
1999	Retired as president.

Their clothing, homes, and social customs resemble those of middle-class North Americans and Europeans. Most white families live in single-family homes, and many employ household help who are not white.

Afrikaners and English-speaking whites generally lead separate lives. They live in different sections of cities, go to different schools, and belong to different churches and social and professional organizations. Afrikaners continue to control most of the nation's agriculture, and English-speaking whites dominate business and industry.

Coloureds

Coloureds make up about 10 percent of South Africa's population. About 85 percent of South Africa's Coloureds live in cities. The Coloured community began in what is now Western Cape, and many Coloureds still live there.

Coloured people have worked for whites for many generations. In cities, many Coloureds have jobs as servants, factory laborers, or craftworkers. In rural areas, many work in vineyards or orchards.

The term *Coloured* is controversial in South Africa. Many Coloured people refer to themselves as *so-called Coloured, classified Coloured,* or *black.*

Asians

Asians make up less than 5 percent of South Africa's population. Their ancestors came mostly from India between 1860 and 1911 to work on sugar plantations.

About 95 percent of South Africa's Asians live in cities. The majority live in Kwa-ZuluNatal. Some have kept many of their old social customs. Most Asians are poor and work in factories or grow vegetables for city markets. But a few are prosperous doctors, industrialists, lawyers, and merchants.

A South African walks on a quiet residential street in Soweto, near Johannesburg. Soweto became a noted tourist destination after the black protests there in 1976.

ECONOMY

The first European farmers arrived in South Africa in the 1650's, and for the next 200 years, the region's economy depended mainly on the raising of livestock and crops. Then in the late 1800's, the situation changed dramatically. Diamonds and then gold were discovered, and mining became the basis of South Africa's economy. Mining started the country on its way to becoming the industrial giant of Africa that it is today.

South Africa produces more gold than any other country, and it ranks as one of the world's chief mining countries. It is a major producer of chromite, coal, copper, diamonds, iron ore, limestone, manganese, phosphate, platinum, uranium, and vanadium.

Gold has been the main force behind South Africa's growth ever since the precious metal was found in the Transvaal in the 1880's. Gold mining has produced great wealth for the country and led to the development of railroads and manufacturing industries.

South Africa's factories today produce almost all the manufactured goods the country needs. Chief products include chemicals, clothing and textiles, iron and steel and other metals, machinery, metal products, motor vehicles, and processed foods. Some South African factories use coal to manufacture gasoline and other fuels.

Most of the nation's factories are in and around the cities of Cape Town, Durban, Johannesburg, Port Elizabeth, and Pretoria. The government offers financial help and other benefits to industrialists who set up new factories near the former black homelands.

Service industries have become increasingly important to South Africa's economy. Service industries include community, government, and personal services, as well as banking, trade, and transportation.

Agriculture remains important to the nation's economy and its people. The leading crops include apples, bananas, corn, grapefruit, grapes, lemons, oranges, potatoes, sugar cane, and wheat. South Africa has an extensive sheep industry, and wool is an important agricultural export. Other leading farm products include beef and dairy cattle, chickens, eggs, milk, and wine.

Electric locomotives wait on the tracks in the railway yard at Durban. South Africa has the best transportation system on the continent, including paved roads that crisscross the nation and six large, well-equipped ports.

Miners toil at a rockface in the Witwatersrand gold field in the Johannesburg area. Hundreds of thousands of people work in the nation's gold mines.

Spraying the vines to keep pests away, a South African farmer protects his grape crop. Wine grapes grown on irrigated land produce famous Cape wines.

South Africa has two main types of farming—that practiced mainly by whites and that practiced mainly by black Africans. White farmers use modern methods and raise products chiefly for the market. Black farm families produce food mainly for their own needs. Black farms are generally much smaller than white farms. Production on black farms has been low because, for many years, blacks were confined to areas where the land was poor. In addition, most black farmers could not afford modern equipment. Since 1994, the government has redistributed some of the country's farmland.

South Africa experienced spectacular economic growth from the 1950's to the 1970's. During these years, many people from other countries invested in South African businesses. In the 1980's, an economic slowdown and international opposition to apartheid led to the withdrawal of some foreign investments. During this time, some countries reduced or ended trade with South Africa.

After the repeal of apartheid in the early 1990's, foreign trade and investment increased. In the 1990's and early 2000's, however, the AIDS disease spread rapidly in South Africa and began to hinder economic growth.

Today, South Africa has Africa's largest economy, producing about 15 percent of the value of the goods and services produced by all African countries combined.

The "Big Hole" at Kimberley, in northern Cape Province, dates from the great diamond rush of the 1070's. The rush brought miners and fortune hunters flocking to the region. South Africa remains famous for its diamonds.

Park rangers oversee animals in the Phinda Private Game Reserve. The reserve covers 56,800 acres (23,000 hectares) of prime wilderness land in KwaZulu-Natal. Phinda is the home of lions, leopards, elephants, buffaloes, and rhinos, as well as more than 400 bird species.

SPAIN

Spain occupies about five-sixths of the Iberian Peninsula, which lies in southwestern Europe between the Atlantic Ocean and the Mediterranean Sea. Portugal, to the west of Spain, occupies the rest of the peninsula.

The Pyrenees Mountains separate Spain from France on the country's northeastern border. These mountains once were a great barrier to overland travel between the Iberian Peninsula and the rest of Europe. Africa lies only about 8 miles (13 kilometers) south of Spain across the Strait of Gibraltar.

Spain's eastern border is the Mediterranean Sea. Spain also includes the Balearic Islands in the Mediterranean Sea and the Canary Islands in the Atlantic Ocean. Madrid, Spain's capital and largest city, stands in the center of the country.

A huge, dry plateau called the *Meseta* covers most of Spain's land area. Poor red or yellowish-brown soil covers most of the Meseta, and little more than scattered shrubs and flowering plants can grow on its plains.

Spain attracts tens of millions of visitors every year. Tourists come to enjoy the country's sunny Mediterranean beaches and islands, as well as its rocky Atlantic coast.

Spain also welcomes visitors to the many castles and churches in its historic cities. The country's long and dramatic history comes vividly to life in the southern city of Granada, which boasts the most famous of the fortified palaces, called *alcazars,* built by the Moors during the Middle Ages.

Northwest of Madrid stands El Escorial, one of the world's largest buildings and an awesome reminder of the Golden Age of Spain. The Golden Age began with the discovery of America by Christopher Columbus in 1492, and it reached its height in the 1500's, as the new territories brought great wealth to Spain.

In later years, wars and the expense of ruling a huge empire caused a decline in Spain's riches. Then, in the 1900's, under the dictatorship of Francisco Franco, Spain became isolated from the rest of European culture. During the 1950's and 1960's, rapid economic development changed Spain into an industrial nation. Spain became a democracy after Franco died in 1975.

Today, Spain is a member of the European Union and uses the bloc's common currency, the euro. However, pride in its achievements has been tempered in the 2000's by its struggles with a severe economic recession.

SPAIN TODAY

When Juan Carlos was crowned king of Spain in 1975, his words rang out across the country: "Today marks the beginning of a new era for Spain." Over the years, Juan Carlos's dream for his country has come true. After 36 years of dictatorship under Francisco Franco, Spain now has a democratic form of government called a *parliamentary monarchy*.

When Franco became dictator in 1939, he took complete control of Spain. His regime was similar to a *fascist dictatorship*—a government in which political choices and personal freedoms were greatly restricted. In 1969, Franco declared that Prince Juan Carlos would become king of Spain after Franco's death or retirement. Franco died in November 1975, and Spain then entered a period of major political change.

From dictatorship to democracy

In 1975, when Juan Carlos became king, Spain was in the midst of terrible internal conflict. Protests and strikes were tearing the country apart. Adding to the general chaos were terrorist attacks by people in the culturally distinct Basque region of northern Spain. These people were demanding independence for their province.

King Juan Carlos I and his supporters quickly began changing Spain's government from a dictatorship to a democracy. First, the new monarchy ended Franco's ban on political parties. In 1977, the government held elections for seats in Spain's parliament, called the *Cortes*. For the first time since 1936, the people of Spain had a choice of candidates in an election.

Then, in 1978, the voters approved a new constitution that was based on democratic principles. The Constitution also gave Spaniards complete religious freedom.

Government leaders continued their reform programs into the 1980's. Censorship was relaxed, and political prisoners were freed. The national government also gave more power to local governments.

FACTS

Official name:	Kingdom of Spain
Capital:	Madrid
Terrain:	Large, flat plateau surrounded by rugged hills; Pyrenees in north
Area:	195,365 mi² (505,992 km²)
Climate:	Temperate; clear, hot summers in interior, more moderate and cloudy along coast; cloudy, cold winters in interior, partly cloudy and cool along coast
Main rivers:	Tagus, Guadiana, Ebro, Duero (Douro), Guadalquivir
Highest elevation:	Pico de Teide, 12,198 ft (3,718 m), in the Canary Islands
Lowest elevation:	Atlantic Ocean, sea level
Form of government:	Parliamentary monarchy
Head of state:	Monarch
Head of government:	Prime minister
Administrative areas:	17 comunidades autonomas (autonomous communities)
Legislature:	Las Cortes Generales (General Courts or National Assembly) consisting of the Senado (Senate) with about 260 members serving four-year terms and the Congreso de los Diputados (Congress of Deputies) with 350 members serving four-year terms
Court system:	Tribunal Supremo (Supreme Court)
Armed forces:	221,800 troops
National holiday:	National Day - October 12 (1492)
Estimated 2010 population:	45,898,000
Population density:	235 persons per mi² (91 per km²)
Population distribution:	77% urban, 23% rural
Life expectancy in years:	Male, 77; female, 83
Doctors per 1,000 people:	3.3
Birth rate per 1,000:	11
Death rate per 1,000:	9
Infant mortality:	4 deaths per 1,000 live births
Age structure:	0-14: 15%; 15-64: 68%; 65 and over: 17%
Internet users per 100 people:	60
Internet code:	.es
Languages spoken:	Castilian Spanish (official), Catalan, Galician, Euskara
Religions:	Roman Catholic (predominant), Protestant, Muslim
Currency:	Euro
Gross domestic product (GDP) in 2008:	$1.612 trillion U.S.
Real annual growth rate (2008):	1.1%
GDP per capita (2008):	$36,068 U.S.
Goods exported:	Automobiles, fruit and vegetables, iron and steel, petroleum products, pharmaceutical products
Goods imported:	Automobiles, chemicals, food, machinery, petroleum and petroleum products
Trading partners:	China, France, Germany, Italy, Netherlands, Portugal, United Kingdom, United States

The political changes in Spain helped create a stronger economy. The quality of food, housing, and health care improved rapidly, and Spaniards now enjoy a higher standard of living than ever before.

Recent developments

In 1982, the Socialist Workers' Party won the most seats in parliament, and Felipe González became prime minister. The elections gave Spain its first leftist government since 1939. González remained prime minister until 1996.

In 1996, the Popular Party won control of parliament, and José María Aznar became prime minister. In 1998, the ETA, a Basque guerrilla group, declared a cease-fire, but it resumed its terrorist attacks in 2000. From 2006 to 2011, the ETA declared several cease-fires. However, the group did not disarm, break up, or renounce its purpose of creating an independent Basque nation.

In 2004, bombs exploded on commuter trains in Madrid, killing nearly 200 people. The government blamed Islamic terrorists for the attacks. In elections held shortly after the attacks, the Socialist Workers' Party won control of parliament, and José Luiz Rodríguez Zapatero became prime minister.

Spain suffered a severe economic recession following a global financial crisis in 2008 and 2009. The Popular Party won the election in 2011, and Mariano Rajoy became prime minister.

ENVIRONMENT

From the sunny Mediterranean coast to the soaring heights of the Pyrenees Mountains, Spain's scenic landscape offers many interesting features. Spain's land regions include the Meseta, the Northern Mountains, the Ebro and Guadalquivir river basins, the Coastal Plains, the Balearic Islands off mainland Spain, and the Canary Islands off the northwest coast of Africa.

The Meseta

The Meseta, a huge, dry plateau that covers most of Spain, consists mainly of broad, flat plains, broken up by hills and low mountains. Higher mountains border the region on the north, east, and south. The Meseta extends into Portugal on the west.

Sheep and goats graze in the highlands of the Meseta, but the region's poor soil makes farming almost impossible. Most of Spain's major rivers rise in the Meseta, and forests blanket the mountains and hills. Throughout the year, the weather is mostly dry and sunny, with hot summers and cold winters. More rain falls in winter than in summer.

The Northern Mountains

The Northern Mountains extend across the north of Spain from the Atlantic Ocean in the west to the Coastal Plains in the east. The Galician and Cantabrian mountain ranges rise sharply from the ocean along most of the Atlantic coast, giving way to the Pyrenees in the east.

The Pyrenees Mountains, which form a natural barrier between Spain and France, cover an area of more than 20,000 square miles (52,000 square kilometers). The mountains rise to an average height of 3,500 feet (1,070 meters), with many peaks in the central ranges soaring more than 10,000 feet (3,000 meters). Because the Pyrenees are so high and cover such a large area,

The snowy peaks of the Sierra Nevada tower over the castle of Calahorra in southern Spain. The Sierra Nevada separates the Andalusia region from the Mediterranean Sea. The range's highest peak, Mulhacén, rises 11,411 feet (3,478 meters).

these mountains tended to isolate Spain from the rest of Europe through much of its history.

River basins

Spain has two major river basins—the Ebro Basin in the northeast and the Guadalquivir Basin in the southwest. The broad plains of the Ebro Basin surround the Ebro River as it flows from the Cantabrian Mountains southeastward to the Mediterranean Sea. The plains of the Guadalquivir Basin spread out along the Guadalquivir River from its source in the Andalusia region to the Gulf of Cádiz on the Atlantic coast.

Although both river basins are very dry, the soil is rich and fertile due to extensive irrigation and reservoirs that store winter rainfall. Major dams on the Ebro near Lérida provide hydroelectric power for the industries of Catalonia.

With a landscape of rolling hills and plains, Andalusia is one of Spain's most scenic regions.

The Cantabrian Mountains form the central part of the range that extends across northern Spain. Thick forests blanket the Cantabrian slopes, and swift-flowing rivers plunge through the mountains. The region has a wet, cool climate.

The Spanish landscape is mostly a dry plateau surrounded by mountains. Only shrubs and small plants grow on the plains. But along the Mediterranean coast, fertile plains provide farmland, and sandy beaches attract tourists.

Coastal Plains

The Coastal Plains, which extend along the entire length of Spain's Mediterranean coast, are the country's richest agricultural area. Farmers along the southeast coast use water from nearby rivers to irrigate their crops of grapes, olives, and oranges and other citrus fruits.

The Coastal Plains' sunny beaches, as well as almost year-round sunshine, attract many visitors and residents to this region of Spain. Resort areas such as Costa Brava, Costa Blanca, and Costa del Sol draw tourists from around the world.

The islands

The Balearic Islands, which lie about 50 to 150 miles (80 to 240 kilometers) east of mainland Spain in the Mediterranean Sea, include five major islands and many smaller ones. The largest of the group is Majorca, a fertile island with a mountain range along its northwest coast. Minorca, the second-largest island, is mostly flat, with wooded hills in the center. Ibiza, the third-largest island, is hilly.

The Canary Islands lie in the Atlantic Ocean about 60 to 270 miles (96 to 432 kilometers) off the northwest coast of Africa. Spain's highest mountain, Pico de Teide, rises 12,198 feet (3,718 meters) in the center of Tenerife, the largest of the Canaries.

Many hotels and villas line the sandy beaches of the Costa Brava on the coast of Catalonia, a popular resort area.

PEOPLE

A drive across the Spanish countryside might lead a traveler to think the land had been deserted long ago. Hundreds of miles of scrub-covered plains give inland Spain a wild and desolate appearance. Occasionally, a small village breaks up the rolling landscape of the Meseta. Whitewashed houses with tile roofs cluster together, surrounded by pastureland. These simple dwellings are a welcome sign of life in an otherwise bleak region.

Inland Spain is very thinly populated because most of the land is difficult to farm. About three-fourths of the Spanish people now live in cities. When the economic development of the 1960's created more jobs in urban areas, the promise of regular employment and higher wages caused many people to abandon their rural villages and head for the cities.

A modern culture

Today, Spaniards who live in the cities have adopted a modern way of life. They live in apartments, wear Western-style clothing, and enjoy a high standard of living. They even live longer than most of their fellow Europeans. Spanish men live an average of 77 years, and the average life expectancy for Spanish women is 83 years.

Even with the trend toward modern ways of life, some age-old customs are still followed. For example, most Spanish factories and businesses close for three hours at lunchtime and then stay open until about 7 p.m. And though the traditional *siesta* (afternoon nap) is disappearing, Spaniards still enjoy the customary *paseo* (walk) before their evening meal.

Regional groups and languages

During the 1200's, the Spanish province of Castile became an important literary, military, and political center. As the influence of the Castilians spread, so did their language. The Castilian dialect soon became the accepted form of Spanish in most parts of the Iberian Peninsula. Today, Castilian Spanish is the official language of Spain.

Seville's oldest restaurant and tavern, El Rinconcillo, was established in 1670. It is noted for tapas, which are Spanish appetizers.

Some regional dialects live on. In the past, many Spanish people felt a greater loyalty to their own region than to the nation as a whole. Three regional groups in northern Spain remain strong today: Basques, Catalonians, and Galicians. Each group is fiercely proud of its own language and traditions. Each group also includes some members who hope for independence from Spain.

Millions of Basques live in the Pyrenees—most on the Spanish side of the mountains. The Basque people speak a language called Euskara or Euskera. The major city in Spain's Basque region is Bilbao, on the Nervión River, near the Bay of Biscay. Bilbao is the capital of Vizcaya province and an important industrial and financial center.

Since the late 1960's, some Basques have been very vocal in their demands for political independence from Spain. Others simply want greater control over their local government affairs. In 1980, the Spanish government granted the Basques limited self-rule.

The people of Catalonia in northeastern Spain are another large minority group. Like the Basques, some want more control over their region's government, and some desire independence. The Catalan language, which is similar to the Provençal tongue of southern France, is widely spoken in Catalonia.

The Catalonians share their history and language with the people of the Balearic Islands. Like the islanders, the Catalonians have been seafaring people for centuries.

A third regional group lives in Galicia, in the northwestern region of Spain. This fairly isolated region was one of the last regions conquered by the Moors and the first to be retaken by Christian forces, both in the A.D. 700's. The cathedral of Santiago de Compostela, said to be the tomb of the apostle James, was one of the three most important pilgrimage centers in western Europe during the Middle Ages. Galicians speak a dialect similar to Portuguese.

Plaza Catalunya lies in the center of Barcelona. Some of the city's major streets and avenues meet at the plaza. The space is especially known for its fountains and statues, and for the flocks of pigeons that gather, seeking food from visitors.

Sunlight casts a warm glow over a woman selling nuts in a Madrid street. Although modern ways have replaced many old customs in the cities, Spanish urban life has kept much of its relaxed quality.

SPANISH FIESTAS

"Every day seems a holiday!" exclaimed Richard Ford, who wrote a *Hand-Book for Travellers to Spain* in 1845. Ford, like many visitors to this sunny country, was impressed by the Spanish love for a *fiesta* (festival). Even today, fiestas are going on all year round and in every part of the country.

Fiestas in different regions of Spain have their own unique character. In Valencia and Catalonia, for example, the fiestas have many bonfires and fireworks. Celebrations in Andalusia feature the colorful, swirling dresses of flamenco dancers, accompanied by the clicking of their handheld *castanets*. But wherever they take place, the high spirits and joyful activity of Spanish fiestas make them among the most exciting in Europe.

Spain's festival year begins in February, with carnivals taking place in several towns throughout the country. In the Andalusian port of Cádiz, the streets fill up with people in brightly colored costumes. During the carnival, they reenact historical events in a humorous way. Because many of these reenactments poke fun at people in authority, they were banned during Franco's dictatorial rule.

Holy Week celebrations

The cheerful liveliness of the February carnivals gives way to the more solemn Christian observances of Ash Wednesday and Lent. *Semana Santa* (Holy Week) is the final week of Lent, which ends in the celebration of Easter. Spain's Holy Week processions are the most famous in all of Europe.

In Seville, the entire city is transformed into a huge, colorful spectacle, as if it were a vast stage with thousands of actors on it. One of the most unique aspects of the Holy Week festivities in this city are the *nazarenos* (penitents).

Dressed in cloaks and pointed hoods, the nazarenos form a procession through the streets, carrying candles and crying out for forgiveness of their sins. They are followed by people carrying *pasos,* elaborately carved religious images mounted on wooden platforms.

The Feria

Holy Week is no sooner over than the Seville *Feria* begins. The Feria is Spain's most famous festival—a weeklong, almost nonstop party that fills the streets with singing, dancing, and celebrating crowds.

The hooded penitents called nazarenos march through the streets of Seville as part of a Holy Week procession. Along the way to the cathedral, the penitents often sing saetas, mournful laments of their sins. Up to 100 processions take place during the festival.

The "running of the bulls" takes place during the Festival of San Fermin in Pamplona every year from July 7 through July 14. The run stretches from the corral where the bulls are kept, to the bull ring where they will fight that same afternoon. Men from all over the world come to Pamplona to risk injury or death by running with the bulls.

Flamenco dancers at the Feria de Cabello in Jerez clap their hands, stamp their feet, and swirl to the music of guitars and castanets. Flamenco dancing and music was first performed by the Gypsies of southern Spain.

↑ Human "castles," rising precariously above cheering crowds, are a regular feature of fiestas in Catalonia. Fireworks, giant puppets, and Catalan music and dance add to the exuberant atmosphere.

The Feria, which traces its origins to a horse and cattle market of the late 1840's, includes street parades, horse races, and bullfights. Women wear brightly colored, polka-dotted costumes with long trains and silk shawls. Decorated booths and tents line the streets, and peddlers sell merrymakers a glass of sherry or cognac.

Pilgrimage to the past

Just a short time after the Feria ends, the pilgrimage to El Rocío begins. In this isolated village about 60 miles (96 kilometers) from Seville, a miraculous image of the Virgin Mary is said to have been seen during the Middle Ages. Pilgrims in colorful costumes travel to El Rocío from all parts of Spain, many on horse-drawn wagons.

The celebrations of the pilgrimage combine the solemn, religious nature of Holy Week with the gaiety of the Feria. Many pilgrims believe that the sand dunes and marshes around the shrine have a spiritual energy.

Spaniards also hold festivals in honor of their local *patron* (guardian) saint. One of the best known of these celebrations is the fiesta of San Fermin, held in July in Pamplona. As part of the festivities, bulls are turned loose in the streets. Young men run ahead of the animals to the bull ring, where bullfights take place.

Bullfighting, a contest between a bull and a *matador* (bullfighter), has a long tradition in Spain, Portugal, southern France, and Mexico. The matador wears a traditional costume decorated with gold thread and sequins, known as a *suit of lights*. In the bull ring, the matador waves a cape or a piece of cloth to maneuver the charging bull, and later uses sticks mounted with steel barbs to stab the animal before it is killed.

In recent years, many people, including Spaniards themselves, have begun to criticize bullfighting as a savage display of cruelty to animals. In 2010, the Catalan parliament voted to outlaw the sport beginning in 2012.

ECONOMY

Until the mid-1900's, Spain was one of the most underdeveloped countries in western Europe. Although it was a nation of farmers, poor soil and a dry climate kept agricultural output low. And, lacking the natural resources to provide raw materials for industry, Spain lagged behind its neighbors in economic growth.

Boom and crisis

Determined to improve Spain's economy, the Spanish government introduced a series of successful development programs in the 1950's and 1960's. During this period, Spain experienced one of the world's highest rates of economic growth. The country more than tripled its annual production of goods and services, due largely to foreign investment and the booming tourist trade.

Today, Spain is an industrial nation and one of the largest economies in the European Union. However, the end of a housing boom in 2007 and a worldwide economic crisis led to high unemployment and a soaring national debt. The government was forced to adopt strict austerity measures to avoid defaulting on its loans.

Tourism

With tens of millions of visitors each year, Spain has one of the world's leading tourist industries. Tourism employs many Spanish workers and brings the country billions of dollars each year in income. The large amounts of foreign currency brought in by tourists help balance the national trade deficit.

Most of Spain's visitors come from other European nations. Prices in Spain, which are lower than in most other countries of western Europe or in the United States, attract many vacationers.

The Spanish government encourages the growth of tourism and closely supervises the quality of accommodations and services offered to tourists. Government-run schools train hotel managers, tour guides, chefs, and other professionals involved in the tourist business.

A nuclear power plant towers over a city in the province of Valencia. With the growth of industrialization in Spain, use of nuclear energy has increased.

The vineyards of La Rioja in northern Spain are noted for their high-quality grapes. The first sherry wines came from grapes grown in Jerez de la Frontera in the southwest. Spain is now one of the world's largest producers of wine.

A flock of sheep grazes on the slopes of the Cantabrian Mountains where open grasslands make ideal pastureland for livestock. Sheep are Spain's most important domestic animals.

Manufacturing

Spain's major automobile plants are located in Barcelona, Madrid, Saragossa, and Valencia. Although the iron and steel industries suffered a decline in the 1980's, iron and steel are still among the nation's most important products.

Other products manufactured in Spain include cement, chemical products, machinery, plastics, rubber goods, ships, shoes and other clothing, and textiles. Madrid is a center for electronics and other high-technology industries.

Agriculture

Agriculture is the weakest part of the Spanish economy, but improvements in irrigation methods and equipment have increased farm production. Spain now ranks among the world's leading producers of olives, oranges, strawberries, and wine. Grain crops, such as barley and wheat, are grown mostly in the northern regions, and grapes, olives, and oranges and other citrus fruits are grown in the south and east.

In the central regions, where much of the land is too poor for agriculture, farmers raise livestock. Beef and dairy cattle, chickens, goats, pigs, and sheep graze on the pastureland.

Spain's reputation as a wine producer continues to grow. Red wine from the country's vineyards has become quite popular.

Spain has overcome its limited natural resources and poor farm-land to develop a stronger, diversified industry-based economy. Manufacturing centers are located mainly in the northern half of the country.

Fruit
Gijón
Oviedo
Potatoes
Titanium
Corn
Hogs
Beef cattle
Lead, Zinc
Coal
Coal
Bilbao
Iron ore
Milk
Sheep
Wheat
Beef cattle
Potatoes
Wheat
Grapes
Potash
Wheat
Sugar beets
Wheat
Sugar beets
Barley
Cork
Valladolid
Barley
Saragossa
Potash
Barcelona
Sheep
Vegetables
Wheat
Silver
Grapes
Goats
Salt
Petroleum
Madrid
Iron ore
Olives
Grapes
Tobacco
Sugar beets
Sheep
Olives
Hogs
Barley
Oranges
Fruit
Olives
Mercury
Grapes
Wheat
Valencia
Sheep
Goats
Fruit
Cork
Grapes
Uranium
Lead
Oranges
Córdoba
Copper
Fruit
Cartagena
Pyrite
Seville
Olives
Goats
Lead, Zinc
Titanium
Rice
Citrus fruits
Cork
Grapes
Málaga
Salt

Irrigated cropland
Other cropland
Mostly grazing land
Forest land
Generally unproductive land
• Manufacturing center
• Mineral deposit

MADRID

Perhaps no other city in Europe can match the energetic, dynamic quality of Madrid. By day, the modern business district of this sprawling metropolis is bustling with activity. By night, the *movida* (young trendsetters) crowd the city's restaurants and nightclubs, dining and enjoying themselves until almost daybreak. On summer evenings, Spaniards of all ages enjoy a leisurely stroll down the Castellana, the city's main boulevard.

With its high-rise apartments, its industrial suburbs, and even its air pollution, Madrid today is much like any other large, modern European city. Yet Madrid is also a historic city, steeped in the past. Its beginnings can be traced to the A.D. 900's, when the Moors built a fortress called Magerit on the site of what is now Madrid. Spanish Christians conquered the area in 1083, and Madrid remained just another small town on the high plains for centuries thereafter.

A statue of King Alfonso XII is surrounded by the formal gardens of the Retiro, a huge park that includes a boating lake and a famous botanical garden. The Retiro dates back to the 1400's and was transformed in the 1630's, when it served as a retreat for Philip IV.

The nation's capital

In 1561, Philip II decided to make Madrid the capital of Spain, mainly because the city is situated almost in the exact geographical center of Spain. Once it became the capital, Madrid grew rapidly into one of the great cities of western Europe.

Soon, many beautiful buildings and plazas transformed the city. Some of these structures still stand, even though much of the city was destroyed during the Spanish Civil War (1936-1939).

Southwest of the *Puerta del Sol* (Gate of the Sun), in the center of downtown Madrid, lies the old section of the city. Some of the buildings along the narrow, winding streets of this area were erected during the 1500's and 1600's. At the western end of the old section stands the Royal Palace, built in the 1700's and the residence of the Spanish royal family until 1931. Many visitors come to see the beautiful palace gardens around the building, which is now a museum.

Center of culture

As the capital of Spain, Madrid also became an important cultural center. In 1605, the first part of Miguel de Cervantes's *Don Quixote* was published in Madrid. Although long thought of as the humorous tale of a madman's adventures, *Don Quixote* was later recognized as an important literary work. It has been a major influence on the development of the novel.

Later in the 1600's, the painter Diego Velázquez, one of the most brilliant artists of the period, moved from Seville to Madrid. Many other artists followed Velázquez's lead, and Madrid soon became a major center of artistic expression.

Today, Madrid's rich cultural heritage can be seen in the Prado, also called the National Museum of Painting and Sculpture. The Prado is one of the world's outstanding art museums, with a collection of thousands of paint-

A large, crescent-shaped plaza, the Puerta del Sol, is the center of downtown Madrid. In the middle of the square is a large equestrian statue of King Carlos III. The king faces the former main post office, built in the 1800's.

The Plaza de España, one of Madrid's largest and most popular squares, is located at the end of the Gran Vía, one of Madrid's busiest streets. It features a large fountain and a towering monument that honors the Spanish writer Miguel de Cervantes.

A student artist uses a Madrid sidewalk to make a chalk copy of Diego Velázquez's masterpiece, *Las Meninas* (The Maids of Honor). The original work, painted in 1656, is displayed with many of Velázquez's other paintings in the Prado, Madrid's famous art museum.

ings by Spanish and foreign artists. Many of the masterpieces at the Prado were originally displayed in the royal collections of Philip II and Charles I.

The early 1900's were an important time in Madrid's cultural life. Many of the city's leading writers and artists became part of the *Residencia de Estudiantes* (Residence of Students), an educational institution that explored new ideas and philosophies. The group included poet Federico García Lorca, painter Salvador Dalí, and film director Luis Buñuel.

The cultural excitement of the early 1900's was brought to an end by the Spanish Civil War in 1933. After the war, much of Madrid lay in ruins, and the city became a sad and gloomy place under Franco's dictatorship. However, in the late 1900's, an economic upturn and government programs to construct modern buildings helped contribute to Madrid's growth. The city has since returned to its former glory.

EARLY HISTORY

People lived in what is now Spain more than 100,000 years ago. These early inhabitants were tribes of hunters and gatherers who had come to the Iberian Peninsula from the coast of North Africa.

About 5,000 years ago, a people known as Iberians occupied much of Spain. The Iberians had an advanced culture. They farmed the land, built villages and towns, and developed great skill as craft workers.

El Cid
(1040?-1099)

Averroës
(1126-1198)

King Ferdinand III of
Castile (1198-1252)

During the 1000's B.C., the Phoenicians sailed from the eastern shores of the Mediterranean to settle on Spain's southern and eastern coasts. About 100 years later, a first wave of Celtic people moved into Spain from the north. Another wave followed about 600 B.C. At about the same time, the Greeks set up trading posts along Spain's east coast.

Armies from the powerful North African city of Carthage conquered Spain in the 400's B.C. Under the leadership of the great Carthaginian general Hannibal, these armies used Spain as a base for their attack on Roman Italy in the 200's B.C.

Roman rule

The Romans eventually drove Hannibal's armies from Spain and began their own conquest of the land. However, it took the mighty Roman army almost 200 years to conquer the stubborn, freedom-loving tribes of Spain. When the Roman army conquered what is now Portugal as well, the Iberian Peninsula was united as a Roman province called *Hispania*.

The Romans constructed many cities and towns in Spain, as well as a vast network of roads that connected these cities. To carry water from the rivers and mountains to dry areas, they built huge aqueducts throughout Spain. The Romans also introduced Latin into the province, forming the basis for the Spanish language.

As Hispania became a leading Roman province, many Romans came there to live. Several of Rome's greatest emperors, including Hadrian and Trajan, were born in Spain.

During the A.D. 400's, the West Roman Empire, which included Hispania, collapsed under the pressure of invading Germanic tribes. One of these tribes, the Visigoths, invaded Spain and, by 573, conquered the Iberian Peninsula.

TIMELINE

c. 3000 B.C.	Iberian tribes occupy what is now Spain.
1000's B.C.	The Phoenicians begin to colonize Spain.
c. 900 B.C.	Celts settle in northern Spain.
c. 600 B.C.	Second Celtic wave of immigration. Greeks set up trading posts on eastern coast.
400's B.C.	The Carthaginians conquer much of Spain.
200's B.C.	Hannibal's army attacks Rome from base in Spain.
218-201 B.C.	Rome defeats Carthage in Second Punic War and begins the conquest of Spain. Over the next 200 years, the Romans complete their conquest of Spain and establish the Roman province of Hispania.
26-25 B.C.	Emperor Augustus defeats some of the last rebellious tribes in Spain.
A.D. 50-80	Roman Emperors Trajan and Hadrian born in Spain.
300's	Christianity is established in Spain.
400's	The Visigoths take Spain from the Romans.
711-718	The Moors conquer almost all of Spain.
900's-1000's	Córdoba is center of rich Islamic culture in Spain.
1000's	Christian kingdoms begin to drive the Moors from Spain. Central Moorish authority crumbles and independent Moorish cities are established.
1094	Castilian hero El Cid captures Valencia from the Moors.
1126	Averroës, famous Islamic philosopher, born in Córdoba.
1212	The Christian army defeats the Moors in the Battle of Las Navas de Tolosa.
1220's	King Ferdinand III consolidates the reconquest of much of the Iberian Peninsula from the Moors.
1280's	The Kingdom of Granada becomes the last stronghold of Moorish rule. The Christian kingdoms of Aragon, Navarre, and Castile control the rest of Spain.

The Visigoths set up an independent Christian kingdom in Spain and tried to reestablish a civilization like that of the Romans. However, fighting among their leaders left them too weak, and it took the Moors from North Africa only seven years to conquer the Visigoths.

Spain under the Moors

The Moorish invasion began in 711, and by 718, only the narrow mountainous region across far northern Spain was free from Moorish rule. The Moors were Muslims—followers of the religion of Islam. The Muslims had a more advanced civilization than most Europeans of the time. They had carefully preserved the writings of the ancient Greek, Roman, and Middle Eastern civilizations, and they had also made great discoveries of their own in mathematics, medicine, and other fields of study. As a result, the Muslims brought a new culture to the people of Spain, even sharing ancient manuscripts with European scholars.

Many of the Spanish people became Muslims during this period. Some Spaniards went on pilgrimages to Mecca, in the Arabian Peninsula. These pilgrimages helped link Islamic Spain to the great centers of Muslim culture, and to the literature, science, art, and architecture that flourished there.

Moorish rule began to weaken during the 1000's. Small Christian kingdoms in northern Spain began to spread to the northeastern Mediterranean coast. Castile became the strongest of these growing Christian kingdoms.

Under the leadership of El Cid, the Castilians led the fight against the Moors. Their struggle became known as the *Reconquista* (Reconquest). In 1212, the Crusaders helped the Christian kingdoms drive the Moors out of Spain, and by the late 1200's, Muslim Spain consisted only of the small Kingdom of Granada.

The Alhambra, the last Moorish palace built in Spain, is known for the beauty of its inner courtyards and fountains. The palace contains some of the finest examples of Moorish art in Europe. Built between 1248 and 1354, the Alhambra stands on a hill overlooking the city of Granada.

The Great Mosque at Córdoba was built as a Muslim house of worship in the 700's. More than 1,000 pillars of granite, onyx, marble, and jasper support its arches. Córdoba was an important center of Moorish art and culture during the 900's.

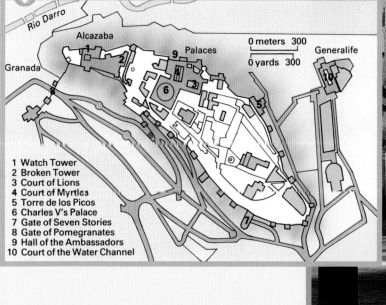

1 Watch Tower
2 Broken Tower
3 Court of Lions
4 Court of Myrtles
5 Torre de los Picos
6 Charles V's Palace
7 Gate of Seven Stories
8 Gate of Pomegranates
9 Hall of the Ambassadors
10 Court of the Water Channel

MODERN HISTORY

By the late 1200's, the small Moorish Kingdom of Granada and the three Christian kingdoms of Castile, Aragon, and Navarre controlled what is now Spain. In 1469, the marriage of Princess Isabella of Castile to Prince Ferdinand of Aragon marked the first step toward a unified Spain. After they inherited their kingdoms, Ferdinand and Isabella together ruled almost all of what was then Spain.

Ferdinand and Isabella thought Muslims and Jews were a threat to their goal of a unified Spain, so in 1480, they set up a religious court called the Spanish Inquisition. The Inquisition, which lasted for more than 300 years, imprisoned or killed people suspected of not following Roman Catholic teachings. Ferdinand and Isabella also began a campaign to drive the Muslims from Granada. In 1492, Granada finally fell to their troops.

The year 1492 also marked a major turning point in history that changed the map of the world forever. In that year, Ferdinand and Isabella sent Christopher Columbus on the voyage that took him to America and brought the first Europeans to the West Indies.

Other Spanish explorers and *conquistadors* (conquerors), such as Vasco Núñez de Balboa, Hernando Cortés, and Francisco Pizarro, followed Columbus to the New World. By the middle 1500's, Spain controlled Central America, Mexico, nearly all the West Indies, parts of North America, and much of South America. In 1512, Ferdinand seized the Kingdom of Navarre, which brought all of what is now Spain under the rule of Ferdinand and Isabella.

El Greco's *View of Toledo* is one of the many great masterpieces created during Spain's Golden Age. El Greco was born on the island of Crete and later settled in Toledo. His paintings had a strong influence on the expressionist painters of the 1900's.

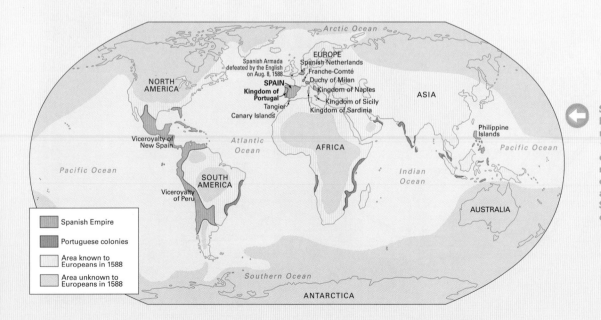

Spanish territories in the New World brought many riches to Spain during the 1500's. But the expense of operating so much territory eventually led to the empire's decline. By 1898, all that remained of Spain's empire were a few outposts in North Africa.

1469	Princess Isabella of Castile marries Prince Ferdinand of Aragon.
1479	The kingdoms of Castile and Aragon are united, bringing almost all of what is now Spain under one rule.
1492	Ferdinand and Isabella's royal forces conquer Granada, the last center of Moorish control in Spain. Christopher Columbus sails to America and claims it for Spain.
1512	King Ferdinand V seizes the Kingdom of Navarre, and all of Spain is united.
1519	Charles I of Spain is crowned Holy Roman Emperor.
1550	Spain gains control of large parts of southern North America, as well as Central and South America.
1556-1598	The Spanish Empire reaches its height—and begins its decline—under the reign of Philip II.
1588	The English navy defeats the Spanish Armada.
1701-1714	The War of the Spanish Succession is touched off by the succession of Philip V to the throne. Spain loses its possessions in Europe.
1808	Napoleon's armies invade Spain and seize Madrid.
1808-1814	Spanish, Portuguese, and British forces drive the French out of Spain in the Peninsular War.
1814	Bourbon King Ferdinand VII returns to the throne of Spain.
1810-1825	All of Spain's American colonies, except Cuba, Puerto Rico, parts of Africa, the Philippines, and Guam declare their independence.
1833	First Carlist War breaks out between those who support the monarchy and those who want a constitutional government.
1874	Republican government is overthrown in the Second Carlist War.
1875	The Spanish monarchy is restored under the reign of King Alfonso XII.
1898	Spain loses Cuba, Guam, Puerto Rico, and the Philippines in the Spanish-American War.
1923-1930	Prime Minister Primo de Rivera rules Spain as dictator.
1931	Spain becomes a democratic republic.
1936-1939	Spanish Civil War brings victory to General Francisco Franco's troops. Franco becomes dictator.
1953	Spain and the United States sign a military and economic agreement.
1975	Franco dies and is succeeded by King Juan Carlos I.
1978	Spain approves a new constitution.
1982	The Socialist Workers' Party, led by Felipe González, wins the most seats in parliament. It remains in power until 1996 elections.
1986	Spain joins the European Community, now known as the European Union.
2004	Terrorists kill 200 in Madrid.
2011	Spain experiences high unemployment and soaring government debt in the wake of a worldwide economic crisis.

The Golden Age

The Spanish Empire reached its greatest heights under the reign of Philip II, who became king in 1556. During this period, Spanish art and literature flourished. Writers such as Miguel de Cervantes, Calderón de la Barca, and Lope de Vega created new literary forms, and Spanish art reached new heights in the paintings of El Greco and Velázquez.

Spain's Golden Age was short-lived, however. In spite of the riches brought from the New World, the country's economy continued to weaken. Huge debts created by a series of costly wars drained the royal treasury. In the 1700's, the War of the Spanish Succession cost Spain all its possessions in Europe. In 1808, Napoleon I conquered Spain, and the French occupied the country until 1814.

Civil war and dictatorship

After the defeat of Napoleon, Spain remained a constitutional monarchy until 1923. In that year, General Miguel Primo de Rivera became prime minister with the powers of a dictator. He was forced to resign in 1930.

Disagreements between Spain's democratically elected government and conservative rebels led to war in 1936. A bloody conflict known as the Spanish Civil War raged for three long years. In 1939, General Francisco Franco and his military forces defeated the forces on the side of the government, and Franco ruled as dictator for the next 36 years. Spain became a democracy after Franco's death in 1975.

King Juan Carlos I succeeded dictator Francisco Franco in 1975. Juan Carlos ended 36 years of dictatorship in Spain by setting up a parliamentary monarchy. His wife became Queen Sophia.

Isabella I
(1451-1504)

Philip II
(1527-1598)

Francisco Franco
(1892-1975)

THE BALEARIC ISLANDS

Off the east coast of Spain in the Mediterranean Sea lie a group of islands whose beautiful scenery, sandy beaches, and mild climate attract millions of tourists each year. They are the Balearic Islands—a province of Spain that includes five major islands and numerous smaller islands.

The main islands of the Balearics are Majorca, Minorca, Ibiza, Formentera, and Cabrera. The city of Palma, on the island of Majorca, is the capital of Majorca and the Balearic province. Ibiza and Formentera are known as the *Pityusae* (Pine Islands). The name comes from the Greek word for a variety of juniper tree. The Balearics are actually the tops of sunken mountain peaks—the eastern part of a range that plunges into the Mediterranean from mainland Spain.

History

The history of the Balearics is a fascinating mixture of ancient legends, romance, and bold adventure. One version claims that the Greek mythological hero Hercules traveled to the Balearics to perform one of 12 labors that would wash away the sins he had committed. Once there, he stole the Golden Apples of the Hesperides from the Tree of Life.

In the 1800's, a writer and artist from Catalonia described the islands as a "lotus land where men are never in a hurry, women never grow old." The beauty and mystery of these islands have drawn many poets, painters, musicians, and philosophers. Composer Frédéric Chopin and novelist George Sand spent time on Majorca.

The islands have been inhabited since prehistoric times. Ancient civilizations left behind huge stone towers, called *talayots,* that still lie scattered throughout the landscape. The talayots, dating from about 1200 to 700 B.C., may have been used for defense. Today, they serve as silent reminders of the Bronze and Iron Age cultures that once flourished on these islands.

The peaceful village life of Majorca, as seen in this cluster of simple homes nestled on a hillside, has inspired artists and writers for centuries. Majorca is the largest of the Balearic Islands. Its name in Spanish is Mallorca.

The carvings on a decorated doorway reflect the sunlight on a street in Ciudadela, the capital of Minorca until 1722. Ciudadela is noted for its shoe industry and fine jewelry.

Sun-drenched beaches and a glittering sea bring millions of tourists to Majorca. The Mediterranean climate of the Balearics—with their hot, dry summers and mild winters—makes these islands a year-round resort. The islands also contain many prehistoric remains, including talayots (round stone towers), navetas (boat-shaped tombs), and taulas (T-shaped stone structures). The city of Palma, on Majorca, has a magnificent Gothic cathedral begun about 1230.

The Carthaginians who eventually settled on the islands were driven out by the Romans in 123 B.C. After Rome fell, the Balearics were held first by the Vandals, and then by the Visigoths, the Byzantines, the Franks, and the Moors. Between 1229 and 1235, James I of Aragon drove the Moors from the islands.

A few years later, James I gave Majorca to his younger son to rule as an independent kingdom. But in 1343, Peter IV reunited the islands with the kingdom of Aragon, on mainland Spain. From that point on, almost without a break, the Balearics belonged to Aragon, and later to the nation of Spain.

Scenic landscape

Despite the high-rise hotels clustered along its beaches today, Majorca still displays much of the natural beauty that has drawn visitors for centuries. Not far from the busy tourist areas are miles and miles of un-spoiled coastline dotted with quiet, secluded bays.

Farther inland on Majorca, rolling plains provide farmland for growing almonds, figs, olives, oranges, and grapes. The plains rise in the northwest to the island's highest elevation—the 4,739-foot (1,445-meter) Puig Mayor. A range of hills called the Sierra de Levante dominates the eastern side.

The quiet, smaller island of Minorca is mostly flat with a rugged coastline. Minorca has felt the influence of tourism much less than Majorca and is best known for its talayots and other prehistoric structures. Farmers on Minorca grow cereals and hemp, as well as grapes, olives, and other fruits.

Ibiza, Formentera, and Cabrera are hilly islands, with many high points dropping in steep cliffs to the sea. Once a hideout for pirates, Ibiza has in recent years become a popular tourist area.

The Balearics include the major islands of Majorca, Minorca, Formentera, Ibiza, and Cabrera, as well as many smaller islands. Most of the smaller islands are little more than rocky islets and are largely uninhabited, though some serve as military and naval bases. Together, the Balearic Islands cover an area of about 1,936 square miles (5,014 square kilometers) and have a population of about 842,000 people. The official language is Castilian Spanish.

An elderly resident of Formentera wears a wide-brimmed hat for protection from the hot sun. Many islanders speak variations of a dialect related to Catalan, a language spoken by people living in the northeastern part of Spain.

THE CANARY ISLANDS

Many millions of years ago, off the northwest coast of Africa, a series of volcanoes erupted from the depths of the Atlantic Ocean. Lava from the repeated eruptions gradually built up, forming mountains on the ocean floor. The tops of these volcanic mountains rising above the ocean are what we know today as the Canary Islands.

Land and climate

In some ways, the landscape of these volcanic islands shows little of their fiery beginnings. The lush tropical vegetation covering most of the islands has created a garden paradise. In other places, however, the lava formations and bowl-shaped craters left by the volcanic activity are vivid reminders of the past.

The Canary Islands make up a group of 13 islands about 60 miles (96 kilometers) off the northwest coast of Africa. Together, the islands cover a total land area of 2,796 square miles (7,242 square kilometers) and make up two provinces of Spain. Six of the islands are uninhabited.

The province of Santa Cruz de Tenerife includes the islands of Tenerife, La Palma, Gomera, and Hierro. The province of Las Palmas includes the islands of Gran Canaria, Lanzarote, and Fuerteventura. Tenerife is the largest of the Canary Islands.

Exploration of these exotic islands began in ancient times. The Romans originally named them *Canaria,* from the Latin word *canis* (dog), because they found large, fierce dogs there.

Pliny the Elder, a Roman who wrote many historical and scientific works, called the Canaries *Fortunate Islands,* or *Islands of the Blest.* Even in modern times, those who have traveled to the Canaries have called them the "islands of eternal spring and autumn" for their near-perfect climate.

The wet, cool breezes of the northeasterly trade winds temper the effects of the hot African

⬆ Bananas, an important ingredient in island cooking, are grown at the higher altitudes of La Palma, where the climate is cooler and more humid than at sea level. La Palma is often called the green island because of its abundant vegetation.

sun. Temperatures range between 68° F (20° C) and 78° F (25° C) and rarely fall below 65° F (18° C) anywhere on the islands. Summers are hot and dry, and winters are warm and wet. When the desert winds from Africa's Sahara reach the Canaries, temperatures rise to about 110° F (45° C).

Many tourists come to the islands seeking relief from winter's cold. The Canaries offer them a peaceful, subtropical paradise of soaring mountains, hidden valleys, rocky cliffs, and lush forests—a perfect refuge from the hectic, modern world.

Desertlike conditions and volcanic soil on the island of Fuerteventura make farming difficult without extensive irrigation.

The Canary Islands lie in the Atlantic Ocean off the northwest coast of Africa. Their location makes them an important refueling point for ships traveling along the West African coast.

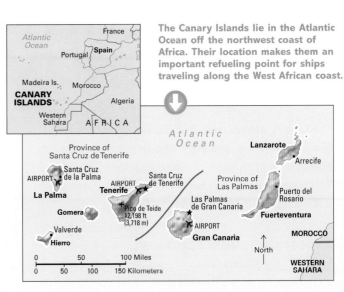

A geyser sprays geothermally heated water from underground with a power like that of the volcanoes that created the Canaries.

Volcanic formations rise from the ground on Tenerife. Pico de Teide, the Canary Islands' highest peak, towers in the background.

Varied landscapes

Steep volcanic ranges, split by deep, dry ravines, cut across all the Canary Islands. On the older islands of Lanzarote and Fuerteventura, the long process of *weathering* (the action of wind and water) has eroded the once-steep mountainsides and formed gently rounded hills. On the western islands, high peaks rise sharply out of the sea, forming steep cliffs.

On Tenerife, the islands' highest mountain—Pico de Teide—towers 12,198 feet (3,718 meters) above sea level. On the island of La Palma, the world's largest volcanic crater, Caldera de Taburiente, measures an astonishing 17 miles (27 kilometers) in diameter. La Palma is covered with immense pine forests. The island of Gran Canaria, with its deserts, mountains, and tropical vegetation, reflects the varied landscape found on all the Canaries.

An abundance of palm trees and wild olive trees are native to the islands. Among the more exotic plants is a dragon tree (*Dracaena draco*), a species that first grew there about 70 million years ago. A small grove of these trees still flourishes on La Palma. The blue-weed (*Echium vulgare*), which blooms against a background of black lava, thrives in the more desertlike conditions of Lanzarote and Fuerteventura.

THE CANARY ISLANDS: HISTORY AND ECONOMY

The Canary Islands were first discovered by ancient Greek and Roman seafarers, who returned to their homes with many tales of these fascinating islands. Yet it was not until the early 1400's that anyone made a serious attempt to conquer the Canaries.

Exploration and conquest

In 1402, the French explorer Jean de Béthencourt led an expedition to the islands, landing first on the north side of Lanzarote. From there, he conquered Fuerteventura and Hierro. Béthencourt received the title *King of the Canary Islands* but recognized King Henry III of Castile, who had provided aid during the conquest, as his overlord.

Béthencourt also established a base on the island of Gomera, but it would be many years before the island was truly conquered. The people of Gomera, as well as the Gran Canaria, Tenerife, and La Palma people, resisted the Spanish invaders for almost a century.

By 1495, the islands had fallen to Spanish rule. The town of Santa Cruz, on La Palma, became a stopping point for Spanish conquerors, traders, and missionaries on their way to the New World.

The islands became very wealthy and soon attracted merchants and adventurers from all over Europe. Magnificent palaces and churches were built on La Palma during this busy, prosperous period. Of particular interest to visitors is the Church of El Salvador, one of the island's finest examples of the architecture of the 1500's.

A proud heritage

When Jean de Béthencourt first arrived in the Canary Islands, he found the Guanches living there. The Guanches were a blond-haired people who lived in caves. They worked with stone tools and carved primitive writing into the rocks.

The whitewashed farmhouses of Lanzarote are similar to those found in North Africa. The terraced hillsides conserve surface water and make farming possible in the dry conditions of the island. Lanzarote is famous for its watermelons and other melons. The island's malmsey wine is made from grapes grown in small craters.

Camels carry
tourists on the trail
to the top of a vol-
cano on Lanzarote.
Modern hotels and
resorts accommo-
date the millions of
vacationers who
flock to the Canary
Islands each year.

A shepherd wearing a
cone-shaped woolen hat
pauses for a rest on the
island of Hierro. The
island is mountainous,
with steep cliffs rising
from the sea. It has large
groves of pine and
beech trees. Much of the
central portion of the
island is a plateau,
reaching a height of
about 4,350 feet (1,330
meters) on Mal Paso.

Present-day islanders trace their heritage to
the Normans, southern Spaniards, and Irish immigrants who intermarried with the original
Guanches. Almost all the islanders are Roman
Catholics.

Today, many age-old island traditions are
giving way to changes brought by new technology. In many respects, however, the islanders have kept the simple life of their
ancestors. To confirm the receipt of goods, for
example, an islander may use the custom of
placing a thumbprint on a ticket. And huge
tanker trucks on their way to make a delivery
must often dodge a team of camels crossing
the road.

Agriculture

The mild climate and fertile soil of the Canary
Islands create ideal farming conditions, though
the dry climate makes irrigation necessary.
Farmers on Lanzarote and Fuerteventura have
developed a unique form of natural irrigation
using porous volcanic stones called *lapilli*. The
lapilli absorb moisture from the trade winds to
produce overnight dew. The dew is then chan-
neled to the roots of the plants.

A variety of crops are grown on the islands,
including apples, avocados, bananas, corn,
lemons, potatoes, sugar cane, tomatoes, and
wine grapes. Tulips and roses are also cultivated, as are date palms and eucalyptus trees.

Because the islands lack natural resources,
many of the people depend on agriculture
and deep-sea fishing for their living. Fishermen catch anchovies, octopuses, and sardines.
The islanders have also responded to the Europeans' growing demands for fresh fruits and
vegetables from the Canaries.

Tourism

Tourism is now the mainstay of the islands'
economy. Many of the islanders work in the
tourist industry, and most of the islands'
gross domestic product comes from tourism.
The islands' sunshine, blue seas, sandy
beaches, and refreshing ocean breezes lure
millions of vacationers every year.

Sri Lanka is a pear-shaped island country about 20 miles (32 kilometers) off the southeast coast of India. Its name means *Resplendent Land,* a reflection of its great natural beauty. Sri Lanka's location, tropical climate, and lush vegetation have inspired many names throughout its history, including *Land of Hyacinths and Rubies, Pearl of the Orient,* and *Teardrop of India.*

The Greeks called the island *Taprobane,* while the seafarers of Old Arabia called it *Serendib.* Many people still know Sri Lanka as *Ceylon,* the country's name from its settlement by Dutch colonists in the 1600's until 1972. The ancient Hindu epic *Ramayana* tells us that "Grey, Green, and Glorious Sri Lanka is like the Garden of the Sky."

Sadly, the island's enchanting scenery cannot hide its many present-day social and economic problems. Sri Lanka suffers from overpopulation, religious and ethnic tensions, a poor economy, and a huge national debt. Its identity as a modern nation often clashes with a strong belief in traditional and religious values.

A landscape of contrasts

Sri Lanka lies in the tropical zone just north of the equator. Mountains rise up to 8,000 feet (2,500 meters) in the south-central region. Sri Lankans grow tea in this region of high mountain walls, narrow gorges, deep valleys, and lofty plateaus. The roads to Nuwara Eliya, which is set on an 8-mile (13-kilometer) plateau in a mountain valley, wind through many of the island's famous tea plantations.

Plains surround the mountains. The land in the southwest is damp and fertile, with a tropical rain forest covering much of the area. The land in the southeast is much drier.

Sri Lanka has warm, even temperatures and high humidity. The low coastal areas average 80° F (27° C). In the mountains, the temperature averages 60° F (16° C). There is very little wind on the island, even though Sri Lanka has two monsoon seasons.

SRI LANKA

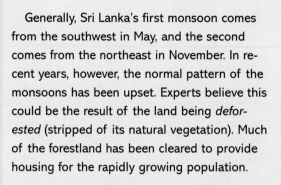

Generally, Sri Lanka's first monsoon comes from the southwest in May, and the second comes from the northeast in November. In recent years, however, the normal pattern of the monsoons has been upset. Experts believe this could be the result of the land being *deforested* (stripped of its natural vegetation). Much of the forestland has been cleared to provide housing for the rapidly growing population.

Only about 35 percent of Sri Lanka's land is still covered by natural tropical vegetation, mostly in nature reserves. The ancient forests of orchids, giant ferns, and teak and ebony trees are now found only in remote mountain regions. However, Sri Lanka's government has become more actively involved in conserving and reforesting the land.

Wildlife

Sri Lankan wildlife is rich and varied. It includes bears, crocodiles, elephants, monkeys, and snakes. Sri Lanka appears to be on a large bird-migration route, and birds from many parts of the world winter on the island.

There is a growing need to protect the island's many rare and endangered species. The clearing of forestland has destroyed the habitats of many birds and animals. In addition, many species have become threatened—or even extinct—because of hunting and trapping.

As early as the 1930's, Sri Lankan officials began setting aside land for the preservation of plant and animal life. Today, strict laws regulate hunting, and much of Sri Lanka's land area is reserved for wildlife protection.

Ruhuna, a famous national park on the southeast side of the island, is home to elephant, deer, and peafowl. Wilpattu, on the northwest side, features many kinds of water birds.

SRI LANKA TODAY

The beauty of present-day Sri Lanka has been marred by tension and fighting between its two largest ethnic groups, the Sinhalese and the Tamils. This fierce and often violent feud began centuries ago. From the A.D. 400's until the arrival of the Portuguese in the 1500's, much of the country's history centered on fighting between Sinhalese and Tamil kings. Because the islanders were so involved in fighting each other, they did little to combat the invasions of European colonists.

The colonial legacy

European control of the country began in the 1500's. The Portuguese arrived in 1505 and soon controlled the island's coastal areas. In 1658, the Dutch gained a monopoly over the island's spice trade and took control away from the Portuguese.

In 1795 and 1796, British troops conquered the Dutch territories. The island, then known as Ceylon, became a crown colony in 1802. The British were the first to gain control of the entire island.

The British established large-scale plantations on Ceylon, and the island soon became a major exporter of coffee, coconut, rubber, and tea. Traditional agriculture, which was based on growing rice to feed the people, was ignored. All efforts were focused on growing the plantations' cash crops.

As a result, Ceylonese villagers were suffering from widespread poverty by the 1900's. Their growing discontent paved the way for independence from British rule. The colony became the independent nation of Ceylon on Feb. 4, 1948.

Recent developments

Ceylon's 1948 constitution was based on the British model, with the queen as head of state. The country adopted a parliamentary form of government headed by a prime minister. In 1972, when a new constitution was adopted, Ceylon was declared a republic and renamed Sri Lanka.

Official name:	Democratic Socialist Republic of Sri Lanka
Capital:	Sri Jayewardenepura Kotte
Terrain:	Mostly low, flat to rolling plain; mountains in south-central interior
Area:	25,332 mi² (65,610 km²)
Climate:	Tropical monsoon; northeast monsoon (December to March); southwest monsoon (June to October)
Main rivers:	Mahaweli, Kala, Aruvi
Highest elevation:	Pidurutalagala, 8,281 ft (2,524 m)
Lowest elevation:	Indian Ocean, sea level
Form of government:	Republic
Head of state:	President
Head of government:	President
Administrative areas:	9 provinces
Legislature:	Parliament with 225 members serving six-year terms
Court system:	Supreme Court
Armed forces:	150,900 troops
National holiday:	Independence Day - February 4 (1948)
Estimated 2010 population:	20,644,000
Population density:	815 persons per mi² (315 per km²)
Population distribution:	85% rural, 15% urban
Life expectancy in years:	Male, 70; female, 76
Doctors per 1,000 people:	0.6
Birth rate per 1,000:	19
Death rate per 1,000:	6
Infant mortality:	17 deaths per 1,000 live births
Age structure:	0-14: 24%; 15-64: 69%; 65 and over: 7%
Internet users per 100 people:	6
Internet code:	.lk
Languages spoken:	Sinhala (official), Tamil (official)
Religions:	Buddhist 69.1%, Muslim 7.6%, Hindu 7.1%, Christian 6.2%, other 10%
Currency:	Sri Lankan rupee
Gross domestic product (GDP) in 2008:	$39.44 billion U.S.
Real annual growth rate (2008):	6.0%
GDP per capita (2008):	$1,958 U.S.
Goods exported:	Diamonds, fish, rubber, tea, textiles and clothes
Goods imported:	Food, machinery, petroleum, textiles, transportation equipment
Trading partners:	China, India, Singapore, United Kingdom, United States

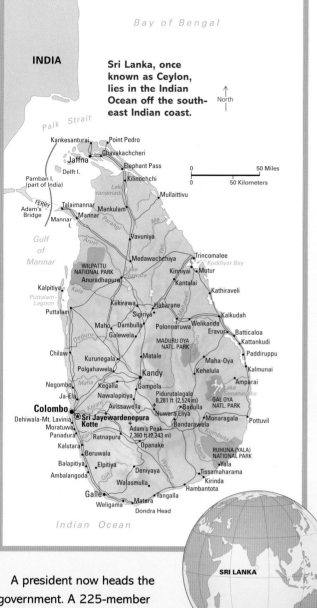

Sri Lanka, once known as Ceylon, lies in the Indian Ocean off the south-east Indian coast.

Fruits and vegetables, including the purple mangosteen shown, make a colorful display at a market in Colombo. In addition to rice and cassava root, Sri Lanka produces a variety of fruits. It also ranks as one of the world's largest producers of tea.

A president now heads the government. A 225-member Parliament passes the nation's laws. The voters elect the president and the members of Parliament to six-year terms.

Sri Lanka faces many challenges. Its rapid rate of population growth puts a great strain on the nation's resources, particularly in education and health care. The population explosion began in the mid-1940's, when a drug to control malaria epidemics became widely used. As a result, the death rate dropped sharply. Government officials hope to control population growth by encouraging people to marry later in life and learn about family planning.

More importantly, the continuing disagreement between the Sinhalese and the Tamils has prevented the people from working together to find solutions to the country's problems. Since 1948, the Sinhalese have controlled the island's government, and they want Sri Lanka to remain a unified state. The Tamils, on the other hand, have sought political independence for the northern and eastern provinces in which they live. Militant Tamil groups—especially the Liberation Tigers of Tamil Eelam (LTTE)—favored the use of force to achieve the separate state. Violence has frequently broken out between government forces and the LTTE, killing tens of thousands of people.

The LTTE abandoned its demand for a separate Tamil state in 2002. In 2008 and 2009, Sri Lanka's military increased its efforts to destroy the LTTE. In May 2009, the military announced it had defeated the rebels.

In December 2004, a tsunami killed about 35,000 people in Sri Lanka, mainly along the island's eastern and southern coasts.

PEOPLE

Sri Lanka's original inhabitants were tribal people called the *Yaksa* and the *Naga*—the ancestors of the *Veddahs*. Most Veddahs have become part of the mainstream culture of Sri Lanka, but a few isolated settlements can still be found in the forests. There, the Veddah people live much as their ancestors did.

The first of the country's invaders came during the 400's B.C. It is believed they were led to the island by the northern Indian prince Vijaya. These early people settled in the northern part of the island and founded the Sinhalese culture.

Ethnic groups

Today, the Sinhalese are the largest ethnic group in Sri Lanka, comprising about 75 percent of the island's population. They speak Sinhala, one of the country's two official languages. Most of them are Buddhists.

The second largest ethnic group are the Tamils. The ancestors of the Tamils arrived on the island from southern India, perhaps as early as the 100's B.C. The Tamils speak Tamil—Sri Lanka's second official language—and most of them are Hindus. They make up about 20 percent of the population.

The Moors, the third largest ethnic group, are the descendants of Arabs and follow the religion of Islam. They make up about 7 percent of the people. Their language is Tamil.

Other ethnic groups include the Burghers and the Malays. The Burghers are descendants of European settlers who intermarried with Sri Lankans. The ancestors of the Malays came to the island from what is now Malaysia.

Religious life

About 70 percent of Sri Lanka's people are Buddhists. As a result, Buddhism forms the basis for spiritual life on the island. The landscape is dotted with many magnificent temples, shrines, and statues. Buddhist monks, known as *bikkhus,* have influenced the island's literary and poetic traditions, as well as

At Lion Rock, the ancient fortress of Sigiriya, all that remains of the brickwork lion is its huge paws. It was built in the A.D. 400's by a ruler named Kasyapa, who killed his father to claim the throne, and then fled to Sigiriya to escape his half-brother.

its art and drama. Because the bikkhus are so highly respected, they also make many high-level political decisions.

Religion has continued to play an important role in the daily life of the Buddhist people. The people hold many religious festivals and ceremonies, including the traditional Poya Days to mark the phases of the moon.

The most important festival of the year for Buddhists is the August festival held at the Temple of the Tooth in Kandy. The temple holds the Buddha's right eyetooth, which was brought to the island in the A.D. 300's. It is the Sri Lankan Buddhists' most sacred relic.

The Temple of the Tooth, in Kandy, holds the right eyetooth of the Buddha. According to Buddhist teachings, the tooth was saved from the funeral pyre. Today, the tooth is kept in an ornate gold casket and guarded closely.

The Perahera festival in Kandy honors the sacred tooth of the Buddha every August. A replica of the casket that holds the tooth is carried through the streets on the back of an elephant while dancers and acrobats entertain the crowds.

Religious tolerance toward those who practice other faiths is an important part of Buddhist teaching. As a result, freedom of worship is enjoyed by people of all faiths in Sri Lanka. The festivals of the four major religions are officially recognized and celebrated as national holidays. In the city of Colombo, Buddhist shrines stand alongside Hindu temples, Christian churches, and Muslim mosques.

It is interesting to note that each faith claims the sacred mountain of *Sri Panda* (Adam's Peak) as its own. Sri Panda rises in the southwest section of Sri Lanka's hill country. At the top of the mountain, there is an indentation in the ground in the shape of a footprint. The Buddhists say that it was made by the Buddha. The Hindus claim it as the god Shiva's. The Christians believe that the footprint was made by St. Thomas the Apostle. The Muslims believe it was made by Adam when he was forced to stand on one foot on the mountain after being sent from Paradise.

Young people worship an image of Ganesh, the Hindu elephant god. Hinduism and Buddhism share many basic features. An image of Ganesh once stood before the Sacred Bo Tree at Anuradhapura and was worshiped by both Buddhists and Hindus.

Sudan is an African country rich in history. Its history is made up of the stories of the individual tribes and regions of Sudan—a mix of peoples as well as landscapes.

Even the name of the land has varied. To the ancient Egyptians, it was *Kush*. To the Greeks, it was part of *Abyssinia,* which some believe came from an Arabic word meaning *mixed*—a reference to the many ethnic groups that lived in the region. To the Romans, it was *Nubia* (from *nub,* which meant *slave* in a local language). And to the Arabs, it was *bilad as sudan (land of the blacks).* In 2011, people in southern Sudan formed an independent nation that they named South Sudan.

As early as the 7000's B.C., prehistoric people lived along the Nile River in what is now Sudan. By about 4000 B.C., the river people had settled in villages to farm and raise animals. Most of these settlers were in northern Sudan, a region that has historical ties with Egypt. The area came to be called Nubia.

Sometime after 2600 B.C., Egypt conquered Nubia. A new civilization developed there after 1000 B.C. that was greatly influenced by Egyptian culture. The Egyptians called the civilization Kush. The Kushites conquered Egypt in 750 B.C. and controlled it for about 80 years. Kush itself existed as an independent kingdom several times, until it collapsed about A.D. 350.

During the 500's, Christian missionaries converted the rulers of southern Egypt and Nubia. But by the mid-600's, Arab Muslims had conquered Egypt and raided Nubia. Later, they made treaties and trade agreements with the Nubians. Arab merchants and religious leaders migrated to the area, and many Arab tribesmen moved to Nubia and married Nubian women.

By the early 1500's, the last of the northern Christian kingdoms had come under Muslim control. The harsh military leaders called *kashifs* who ran the former kingdoms were mainly concerned with tax collecting and slave trading.

During the 1500's, black Muslims called Funj conquered much of what is now Sudan. The Funj sultan of Sennar ruled the tribes and other local groups in the area. Meanwhile, several other black African groups settled the southern part of what are now Sudan and South Sudan. These people

SUDAN

included the Dinka, the Shilluk, the Nuer, and the Azande.

In 1820, the ruler of Egypt sent his son and 4,000 troops to attack and conquer Sudan. By 1821, the kashifs and Funj were under Egyptian control. Eventually, the Egyptians gained control of all of Sudan. Thousands of Sudanese were enslaved.

In 1881, a Sudanese Muslim religious teacher named Muhammad Ahmad proclaimed himself the *Mahdi*—a guide appointed by God. Over the next four years, Ahmad led a successful revolt against the Egyptians. But in 1898, the United Kingdom and Egypt joined forces to defeat the Sudanese at the Battle of Omdurman. The two countries agreed to rule Sudan together, but most of the important officials were British.

During the 1900's, many Sudanese demanded an end to British rule. Some wanted Egyptian control, while others sought independence. The nationalist movement grew in the 1940's and 1950's, and finally the United Kingdom and Egypt agreed to steps leading to self-government. Sudan became an independent nation on Jan. 1, 1956.

The differences in ethnic background, religion, and language between the northern Sudanese and the southern Sudanese troubled the new nation. People in the south were fearful that the northern Muslims would force their beliefs on the whole country.

In 1958, the government was overthrown by General Ibrahim Abboud. In 1964, Abboud in turn was forced out as a result of a general strike by teachers, students, lawyers, and union organizers. Five years later, however, another army officer, Colonel Gaafar Nimeiri, seized power. In 1983, Nimeiri established Islamic law throughout Sudan and ended the regional government in the south.

From 1983 to 2004, about 2 million people were killed in the fighting between Sudan's government and rebels in the south. Hundreds of thousands more died from hunger, as fighting interfered with the production and distribution of food and repeated droughts contributed to the spread of disease.

The Sudanese government and the rebels signed a peace agreement in 2005. In 2011, people in the south voted overwhelmingly to form an independent nation, which they named South Sudan.

SUDAN TODAY

Sudan has always been a land of widely differing geography and a divided people. Vast deserts sprawl across the northern part of Sudan, and grassy plains blanket what was once the central part of the country and is now the southern part. Steamy jungles and swamps lie in the area that became South Sudan. Most Sudanese consider themselves Arabs, speak Arabic, and are Muslims. The people of South Sudan belong to various black ethnic groups, speak any one of a number of languages, and follow African religions or Christianity.

Sudan became an independent nation in 1956, but cultural and social divisions made it difficult to develop a feeling of national unity. Those divisions led to the creation of the new nation of South Sudan in 2011.

Problems between the north and the south

When Colonel Gaafar Nimeiri seized control of Sudan's government in 1969, he agreed that the south would have its own regional government. However, in 1983, he ended this government and established Islamic law—the religious and social law of Muslims—throughout Sudan. Fighting broke out in the south, where few people were Muslims, and a southern guerrilla group called the Sudan People's Liberation Party attacked government buildings.

In 1985, a group of army officers forced Nimeiri out. Their leader, General Abdul Rahman Suwar El-Dahab, abolished the national legislature and set up a military government. However, he soon promised to hold elections for a civilian government.

In 1986, a legislature was elected and began working out a new constitution for Sudan. But in 1989, the military took control of Sudan and suspended the constitution. In 1998, Sudan adopted a new constitution that allowed the formation of political parties. In the late 1990's, Sudan also began cooperating with international companies to develop oil reserves in the southern part of the country. Sudan began to export oil in 1999, and oil soon became the country's leading export.

From 1983 to 2004, Sudan was engulfed in a bitter civil war. The fighting interfered with the production and distribution of food and caused widespread hunger. Drought conditions contributed to the spread of hunger and disease. In 2002, the government agreed to eventually allow southerners to hold a referendum on independence. In 2005, the two sides signed a full peace agreement to end the conflict.

FACTS

Official name:	Jumhuriyat as-Sudan (Republic of the Sudan)
Capital:	Khartoum
Terrain:	Generally flat, featureless plain and desert; mountains in east and west
Area:	718,723 mi² (1,861,484 km²)
Climate:	Hot, dry desert in the north; grass-covered plains in the south
Main rivers:	Nile (Nile, Blue Nile, White Nile), Atbara
Highest elevation:	Jebel Marra, 10,131 ft (3,088 m)
Lowest elevation:	Red Sea, sea level
Form of government	Republic
Head of state:	President
Head of government:	President
Administrative areas:	15 wilayat (states)
Legislature:	National Legislature consisting of the Council of States and a National Assembly
Court system:	Constitutional Court, National Supreme Court, National Courts of Appeal
Armed forces:	N/A
National holiday:	Independence Day - January 1 (1956)
Estimated 2010 population:	36,024,000
Population density:	50 persons per mi² (19 per km²)
Population distribution:	N/A
Life expectancy in years:	N/A
Doctors per 1,000 people:	N/A
Birth rate per 1,000:	N/A
Death rate per 1,000:	N/A
Infant mortality:	N/A
Age structure:	N/A
Internet users per 100 people:	N/A
Internet code:	.sd
Languages spoken:	Arabic (official), English (official), Dinka and other African languages
Religions:	Sunni Muslim (predominant); also traditional African religions and Christianity
Currency:	Sudanese pound
Gross domestic product (GDP):	N/A
Real annual growth rate:	N/A
GDP per capita:	N/A
Goods exported:	N/A
Goods imported:	N/A
Trading partners:	N/A

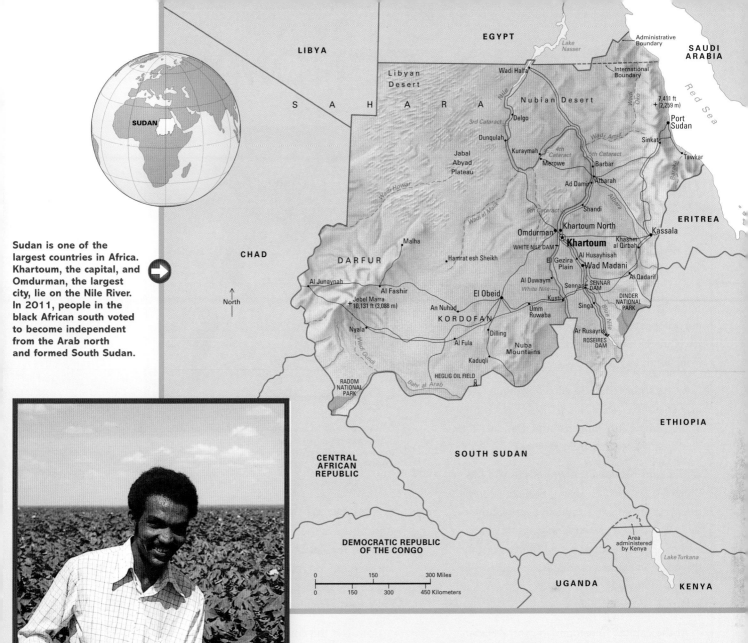

Sudan is one of the largest countries in Africa. Khartoum, the capital, and Omdurman, the largest city, lie on the Nile River. In 2011, people in the black African south voted to become independent from the Arab north and formed South Sudan.

Cotton fields flourish in Sudan's El Gezira region, between the White Nile and Blue Nile. El Gezira is irrigated with Nile water. The Sudanese government controls most of the irrigated farmland in this developing nation.

In a referendum on independence held in the south in January 2011, more than 99 percent of southerners voted in favor of withdrawing from the north. South Sudan became an independent nation in July 2011. The government of Sudan began drafting a new constitution to replace the interim document that had been in effect since 2005. Nevertheless, violent clashes between Sudan and South Sudan continued, primarily over border demarcation and oil revenue issues.

Unrest also continued in western Sudan. In 2003, a separate conflict had erupted in Darfur. Rebels who claimed that the Sudanese government ignored Darfur began attacking government targets in the region. In response, the government and government-backed Arab militias launched attacks in Darfur. More than 200,000 people have died, and millions have been forced from their homes. Many international groups have accused the government and government-backed forces of massive human rights abuses. The United Nations has been in charge of peacekeeping operations in Darfur since 2007.

LAND AND ECONOMY

Before it was divided, Sudan was the largest country in Africa. Now it is the second largest, after Algeria. Sudan sprawls over more than 718,000 square miles (1.8 million square kilometers) of desert and grassland.

The northeastern corner of Sudan borders the Red Sea. The Nuba Mountains rise in the south, and highlands lie in the west and along the country's southeastern border. Southern Sudan is largely a grassy plain. But by far the most important geographic feature in the country is the Nile River, which flows through Sudan from south to north.

In the south, the Nile enters Sudan as two branches. The White Nile enters from South Sudan, where it is known as the Bahr al Jabal. The Blue Nile enters from Ethiopia.

The land between the White Nile and the Blue Nile is called El Gezira. El Gezira contains the most fertile and productive farmland in Sudan. The area has enough water for farming and temperatures that are a little cooler than the north in summer and a little warmer than the north in winter. The rest of southern Sudan is largely a grassy plain.

At the capital, Khartoum, in central Sudan, the White Nile is joined from the east by the Blue Nile. North of Khartoum, the White and Blue Nile merge to form the great Nile River that flows through the desert of northern Sudan. The Libyan Desert to the northwest and the Nubian Desert to the northeast are both part of the Sahara. Rainfall in this desert area rarely amounts to more than 4 inches (10 centimeters) a year. Summer high temperatures average 110 °F (43 °C) and rise as high as 125 °F (52 °C).

Much of the desert north is uninhabited. A few nomads roam the southern edge of the desert and the Red Sea Hills to the east, but most people in the north live along the Nile or the seacoast.

Cotton is Sudan's leading crop and one of its most important exports. The cultivation of cotton was introduced by the British in the early 1900's to supply textile mills in Britain.

A nation of farmers

Agriculture is the mainstay of Sudan's economy. Most of the nation's workers are employed in agriculture, and much of the total value of Sudan's economic production comes from the crops and animals the people raise.

Cotton is the country's leading crop. The Sennar Dam on the Blue Nile provides water to irrigate the cotton fields of El Gezira. Sudanese farmers also grow millet, peanuts, sesame, sorghum, sugar cane, and wheat. *Gum arabic*, harvested from the sap of the acacia tree, is used to make glue.

Crops are also grown in the western highlands, in the Nuba Mountains along the White Nile, and on an irrigated clay plain in the east. A large system of dams and pumps provides irrigation mostly for government projects.

Livestock is an important part of the country's rural economy. Cattle, sheep, goats, and camels provide food as well as income for herders and nomads. Live animals are also exported.

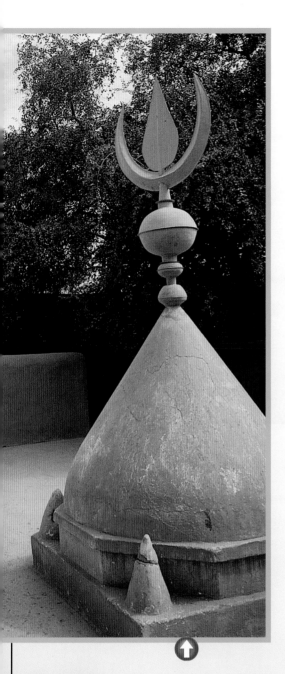

A lone traveler on a mule makes his way along the bank of the Nile River, using an age-old means of transportation. Few of Sudan's people own automobiles.

A muezzin (crier) high atop a minaret calls the Muslim faithful to prayer. The tomb of Muhammad Ahmad, a religious leader known as the Mahdi (divinely appointed guide) appears in the background.

A street scene in Khartoum, the capital of Sudan, shows the country's blend of the modern and the traditional. Khartoum is a center of trade and communications.

Manufacturing and mining

As a developing nation, Sudan is not heavily industrial. Only a small number of the nation's workers are employed in manufacturing and mining. Sudanese factory workers produce food products, petroleum products, shoes, and textiles. Most factories are located in the Nile Valley, especially around Khartoum and in El Gezira. Port Sudan, on the Red Sea, also has some industrial plants.

The leading minerals mined in Sudan are chromium, gold, and gypsum. Geologists have also located deposits of copper, iron ore, lead, nickel, silver, tungsten, and zinc. Dams on the Nile provide hydroelectric power for people in some rural areas. People in other areas use wood from Sudan's forests for fuel.

About three-quarters of Sudan's petroleum deposits were located in the south, in an area that became part of South Sudan. However, most processing facilities are in the north, as are the pipelines that lead to the ports. South Sudan is landlocked. Since South Sudan declared independence, conflict between the two nations has erupted over how to share oil revenues and over possession of some of the oil fields.

PEOPLE

The people of Sudan are one of the most diverse groups in Africa. They belong to many different ethnic groups and speak many different tribal languages. Most people speak Arabic, and English is also spoken. Arabic and English are the nation's official languages.

Most of Sudan's people consider themselves Arabs. Some are descended from Arab immigrants, while others belong to ethnic groups that gradually adopted the Arabic language. Other ethnic groups include Nubians, Beja, Fur, and descendants of West African immigrants.

Most Sudanese are Muslims—followers of the Islamic religion, which Arabs brought to the region many centuries ago. Most are Sunni Muslims. They belong to the larger of the two divisions of Islam. A smaller number practice Christianity or a traditional African religion.

The majority of the Sudanese live in rural areas. Most work as farmers or herders along the Nile River and its two main branches, the White Nile and the Blue Nile.

Most farmers own small plots of land and use old-fashioned tools and farming methods. Many struggle to grow enough food for their families. A small number of rural Sudanese are nomads, wandering the desert with their herds in search of pasture and water.

In the north, many rural Sudanese live in flat-roofed, rectangular houses made of sun-dried mud brick. People in the south build thatch-roofed huts.

Sudan's largest urban center is made up of three cities—Khartoum, Khartoum North, and Omdurman. It is located in the area where the White and Blue Nile come together. Other large cities include Port Sudan, on the Red Sea; Wad Madani, on the Blue Nile, southeast of Khartoum; and El Obeid, in the center of Sudan.

City dwellers generally work in stores, offices, and factories, and many live in apartment buildings or small houses much like those found in Western cities. However, a large number of urban Sudanese are unemployed and live in poor urban neighborhoods.

Workers sort freshly harvested dates in the desert village of Soleib in northern Sudan. Most Sudanese live in rural areas and work as farmers or herders.

Sudan is one of the most diverse countries in Africa. its people belong to a number of different tribal groups and speak many different languages.

A Sudanese woman sleeps in a refugee camp in northern Darfur. Beginning in 2003, ethnic violence in the region resulted in the deaths of tens of thousands of people and the displacement of millions more.

The Kawahla are nomadic camel breeders who move their herds, following the rains. They tend to concentrate in the northern interior of the country.

There, houses are much like those in rural areas. Many people on the outskirts of the major cities live in shantytowns made up of tents and other makeshift shelters.

In the cities, people wear both traditional and Western-style clothing. Most women who wear modern clothing also wear a traditional outer garment called a *taub* that covers them from head to foot. Many men wear a long robe called a *jallabiyah* along with a skullcap called a *taqiyah* or a white turban called an *imamah*.

Most Sudanese do not eat much meat. The main dish throughout Sudan is *ful*—beans cooked in oil and spices. The national drink is *karkadai,* made from the hibiscus plant. The Sudanese also enjoy tea and coffee.

The government provides two years of preschool and eight years of elementary education for free. Students may attend secondary school for three years. The country also has a number of private and public universities, many of which are in

Khartoum. More than half of Sudanese adults can read and write.

Most people have little time for recreation. Still, soccer is the country's most popular sport. Sudanese also enjoy visiting with family and friends. Traditional handicrafts are the most common art form.

SUDAN, SOUTH

South Sudan is a country in eastern Africa. The nation became independent in 2011, when it separated from Sudan. Juba is South Sudan's capital and largest city. It lies on the White Nile River, the country's most important geographic feature. Most South Sudanese live near the river or one of its branches. South Sudan's landscape includes plains, jungles, swamps, and low mountains.

South Sudan is home to many ethnic groups. The Dinka (Jaang or Monyjaan in Dinka) make up the largest group. Other ethnic groups include the Nuer, Shilluk, and Azande. While people of South Sudan speak many different languages, the majority speak either Dinka or Arabic. English serves as the main language of trade, government, and education. Most South Sudanese people follow traditional African religions, but Christianity and Islam are also practiced.

South Sudan is rich in natural resources, including significant oil deposits. The government relies almost entirely on oil for its income. However, South Sudan is landlocked and must transport its oil through pipelines to ports in Sudan. After independence, conflict arose between the two nations over how to share oil revenues.

South Sudan is one of the world's least developed nations. Subsistence agriculture dominates the economy. Most of the country's people farm, fish, or herd animals for a living, but many farmers struggle to grow enough food for their families. Many people lack food, education, or medical care, and the infant mortality rate is one of the highest in the world. Officials hope money from oil reserves will eventually improve living conditions.

Most of the people live in rural areas, many of them in thatch-roofed shelters. Most city dwellers live in apartment buildings or small houses.

FACTS

Official name:	Republic of South Sudan
Capital:	Juba
Terrain:	Plains, jungles, swamps, and low mountains
Area:	248,777 mi² (644,329 km²)
Climate:	Hot and humid with ample rainfall
Main rivers:	White Nile (Bahr al Jabel)
Highest elevation:	Mount Kinyeti, 10,456 ft (3,187 m)
Lowest elevation:	White Nile at northern border, 1,270 ft (387 m)
Form of government:	Republic
Head of state:	President
Head of government:	President
Administrative areas:	10 states
Legislature:	National Legislature consisting of the Council of States, whose members are elected by state legislatures, and the National Assembly, whose members are appointed
Court system:	N/A
Armed forces:	N/A
National holiday:	Independence Day - July 9 (2011)
Estimated 2010 population:	9,100,000
Population density:	37 persons per mi² (14 per km²)
Population distribution:	N/A
Life expectancy in years:	N/A
Doctors per 1,000 people:	N/A
Birth rate per 1,000:	N/A
Death rate per 1,000:	N/A
Infant mortality:	N/A
Age structure:	N/A
Internet users per 100 people:	N/A
Internet code:	.ss
Languages spoken:	Arabic (official), English (official), Dinka and other African languages
Religions:	Indigenous beliefs, Christianity, Islam
Currency:	South Sudan pound
Gross domestic product	N/A
Real annual growth rate (2008):	N/A
GDP per capita (2008):	N/A
Goods exported:	N/A
Goods imported:	N/A
Trading partners:	N/A

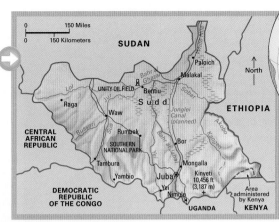

South Sudan is a country in eastern Africa. The nation was formed in 2011 when it seceded (separated) from the rest of Sudan. Juba, which lies on the White Nile River, is the capital and largest city of South Sudan. The White Nile is also called Bahr al Jabel.

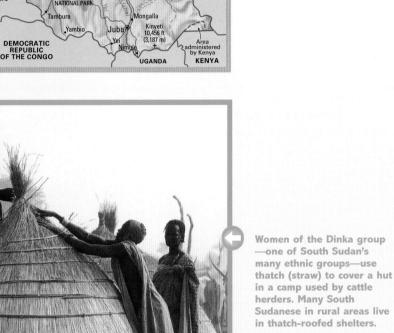

Women of the Dinka group —one of South Sudan's many ethnic groups—use thatch (straw) to cover a hut in a camp used by cattle herders. Many South Sudanese in rural areas live in thatch-roofed shelters.

River transportation is important in South Sudan. Most roads are unpaved, and few of the people own cars. Buses and trucks link cities and towns. Juba and a number of towns have small airports.

People have lived in what is now South Sudan for thousands of years. Ancient kingdoms flourished there, and Egypt controlled parts of the area at various times. Egypt and the United Kingdom ruled Sudan from 1899. Sudan became an independent nation in 1956. However, religious and ethnic differences triggered violence in Sudan for many years.

In 2005, as part of a peace agreement between Sudan's government and rebels in the south, the country's National Assembly approved an interim (temporary) constitution. It provided for power to be shared between northern Sudan and southern Sudan. The constitution also established a regional government in southern Sudan. The regional government had a large degree of self-rule. In January 2011, the people in southern Sudan voted to secede (separate) from Sudan. The new Republic of South Sudan became independent on July 9, 2011.

SURINAME

The Republic of Suriname (also spelled *Suri-nam*) is the smallest independent country in South America, both in area and in population. It is situated on the northern coast of the continent.

Mountainous rain forests cover most of the land in Suriname. An area of swampy flatland that has been drained for farming lies along the northern Atlantic coast. Farther inland, a sandy plain rises about 150 feet (46 meters), and a high grassy *savanna* (treeless plain) stretches along the southwest border.

History

Christopher Columbus sighted the territory that is now Suriname in 1498, but it was not until 1651 that the British built the first permanent settlement there. In 1667, the British gave the territory to the Dutch in exchange for New Amsterdam, which later became the state of New York.

In their own country, the Dutch had learned how to drain low-lying coastal land, and they soon put their experience to work in their new colony. They cleared and drained the swamps along the Atlantic coast and established plantation farms. The Dutch imported African slaves, who worked on these plantations until 1863, when slavery was abolished in the colony. Laborers were then brought from India and Indonesia to work in the fields.

Suriname became a self-governing Dutch territory in 1954 and gained complete independence in 1975. During the 1980's, both civilians and the military held power in the government. Today, the nation is ruled by a civilian democratic government.

People

Many different ethnic groups live in Suriname, and each has preserved its own culture, religion, and language. The Hindustanis, descend-

FACTS

Official name:	Republiek Suriname (Republic of Suriname)
Capital:	Paramaribo
Terrain:	Mostly rolling hills; narrow coastal plain with swamps
Area:	63,251 mi² (163,820 km²)
Climate:	Tropical; moderated by trade winds
Main rivers:	Courantyne, Coppename, Saramacca, Suriname
Highest elevation:	Mount Juliana Top, 4,200 ft (1,280 m)
Lowest elevation:	Atlantic Ocean, sea level
Form of government:	Constitutional democracy
Head of state:	President
Head of government:	President
Administrative areas:	10 distrikten (districts)
Legislature:	National Assemblee (National Assembly) with 51 members serving five-year terms
Court system:	Court of Justice
Armed forces:	1,900 troops
National holiday:	Independence Day - November 25 (1975)
Estimated 2010 population:	466,000
Population density:	7 persons per mi² (3 per km²)
Population distribution:	75% urban, 25% rural
Life expectancy in years:	Male, 69; female, 75
Doctors per 1,000 people:	0.5
Birth rate per 1,000:	17
Death rate per 1,000:	6
Infant mortality:	18 deaths per 1,000 live births
Age structure:	0-14: 28%; 15-64: 65%; 65 and over: 7%
Internet users per 100 people:	9
Internet code:	.sr
Languages spoken:	Dutch (official), English, Sranan Tongo, Hindi, Javanese
Religions:	Hindu 27.4%, Protestant 25.2%, Roman Catholic 22.8%, Muslim 19.6%, indigenous beliefs 5%
Currency:	Suriname dollar
Gross domestic product (GDP) in 2008:	$2.98 billion U.S.
Real annual growth rate (2008):	6.0%
GDP per capita (2008):	$6,515 U.S.
Goods exported:	Aluminum, crude oil, fish, gold, rice
Goods imported:	Food, machinery, petroleum products, transportation equipment
Trading partners:	Canada, Netherlands, Norway, Trinidad and Tobago, United States

Tropical rain forests extend over most of Suriname. They yield a large supply of hardwoods from which Suriname's timber industry produces logs and plywood.

ants of people from India, are the largest ethnic group. They make up more than a third of the population. Some Hindustanis own small farms, while others are industrial workers.

About another third of the people are Creoles—people of mixed European and black African ancestry—who work primarily in business and government. Other groups include Indonesians, Maroons, American Indians, Chinese, and Europeans. Most Indonesians are tenant farmers who rent their land from large landowners.

The variety of languages spoken in Suriname reflects the nation's ethnic mix. Dutch is the official language, but the most commonly used tongue is Sranan Tongo, also called *Taki-Taki*. Sranan Tongo is a mixture of English, Dutch, and several African languages. English serves as the language of business and commerce.

Maroons are the descendants of Black Africans who escaped from slavery. Most live in the rain forests and still follow African tribal customs.

Suriname was known as Dutch Guiana before it became independent. Most of Suriname's people live on the fertile coastal plains. Paramaribo—Suriname's capital, largest city, and chief seaport—has nearly half of the country's population. Suriname's economy is based on mining and metal processing.

SWAZILAND

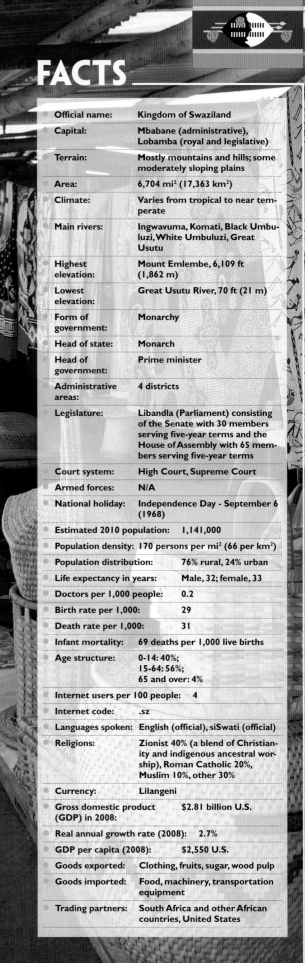

Swaziland is a tiny country in southern Africa. It is surrounded by the Republic of South Africa on three sides and by Mozambique on the east. Swaziland has rich mineral deposits, large forests, and good farmland. The vast majority of people belong to the Swazi ethnic group. However, Europeans own most of the mines, processing plants, and profitable farms.

Most Swazi people are rural farmers and herders. They grow some cash crops as well as food for their families, and they also raise cattle. Swazi farmers prize their cattle, and they respect people with large herds. Traditionally, Swazi do not kill cattle for food. Some cattle are sold for cash or sacrificed at religious ceremonies. When a Swazi man marries, his family gives cattle to his wife's family to make the marriage official.

Swaziland is a landlocked country with mountains on its western and eastern borders. It is one of the best-watered areas in southern Africa. Four main rivers—the Ingwavuma, Komati, Umbuluzi, and Great Usutu—supply the water needed to irrigate crops and to run hydroelectric power plants.

According to the legends of the Swazi, their ancestors lived in what is now Mozambique. In the late 1700's, the Swazi chief Ngwane II led his people over the mountains to what is now southeastern Swaziland. Ngwane II and the chiefs who ruled after him united the Swazi with several other African peoples living in the region.

In the 1830's, British traders and South African farmers of Dutch descent called *Boers* came to Swaziland. By the 1880's, gold had been discovered, and hundreds of prospectors rushed into the region. They asked the Swazi leaders to sign documents granting them rights to mine, farm, and graze animals on the land. Because the Swazi could not read the documents, they did not realize they were giving away their land.

First the South African Boer Republic and later the United Kingdom took control of Swaziland. In 1968, the United Kingdom gave the region its independence.

FACTS

Official name:	Kingdom of Swaziland
Capital:	Mbabane (administrative), Lobamba (royal and legislative)
Terrain:	Mostly mountains and hills; some moderately sloping plains
Area:	6,704 mi² (17,363 km²)
Climate:	Varies from tropical to near temperate
Main rivers:	Ingwavuma, Komati, Black Umbuluzi, White Umbuluzi, Great Usutu
Highest elevation:	Mount Emlembe, 6,109 ft (1,862 m)
Lowest elevation:	Great Usutu River, 70 ft (21 m)
Form of government:	Monarchy
Head of state:	Monarch
Head of government:	Prime minister
Administrative areas:	4 districts
Legislature:	Libandla (Parliament) consisting of the Senate with 30 members serving five-year terms and the House of Assembly with 65 members serving five-year terms
Court system:	High Court, Supreme Court
Armed forces:	N/A
National holiday:	Independence Day - September 6 (1968)
Estimated 2010 population:	1,141,000
Population density:	170 persons per mi² (66 per km²)
Population distribution:	76% rural, 24% urban
Life expectancy in years:	Male, 32; female, 33
Doctors per 1,000 people:	0.2
Birth rate per 1,000:	29
Death rate per 1,000:	31
Infant mortality:	69 deaths per 1,000 live births
Age structure:	0-14: 40%; 15-64: 56%; 65 and over: 4%
Internet users per 100 people:	4
Internet code:	.sz
Languages spoken:	English (official), siSwati (official)
Religions:	Zionist 40% (a blend of Christianity and indigenous ancestral worship), Roman Catholic 20%, Muslim 10%, other 30%
Currency:	Lilangeni
Gross domestic product (GDP) in 2008:	$2.81 billion U.S.
Real annual growth rate (2008):	2.7%
GDP per capita (2008):	$2,550 U.S.
Goods exported:	Clothing, fruits, sugar, wood pulp
Goods imported:	Food, machinery, transportation equipment
Trading partners:	South Africa and other African countries, United States

At the time of independence, the British introduced a constitution that made Swaziland a constitutional monarchy headed by King Sobhuza II. But many Swazi felt the constitution disregarded their traditions and interests.

In 1973, King Sobhuza abolished the constitution, suspended the legislature, and began to rule the country with the aid of ministers. A new constitution, more in keeping with Swazi traditions, was written, and a new legislature was established in 1979.

King Sobhuza died in 1982 after reigning for 61 years. He was succeeded by one of his 67 sons, the 15-year-old Makhosetive, who became King Mswati III in 1986. In 2005, King Mswati approved a new constitution for Swaziland but maintained his power.

Today, the Swazi king, or Ngwenyama, continues to rule Swaziland with the assistance of a council of ministers and a legislature. The queen mother, or Ndlovukazi, is in charge of national ceremonies.

Many Europeans and *Eurafricans* (people of mixed European and African descent) live in Swaziland. The Europeans own farms, mines, and forests and raise many cash crops, including citrus fruits, cotton, pineapples, rice, sugar cane, and tobacco. They also raise cattle for meat, skins, and hides.

Since the 1940's, European companies have planted the highlands of Swaziland with pine and eucalyptus trees, and now the country has one of the largest artificially created forests in Africa. European-owned mills process wood pulp and other forest products.

Rich mineral deposits lie in the mountains, and much of Swaziland's income comes from the European-owned mining industry. The country has deposits of asbestos, barite, coal, gold, kaolin, and tin.

Rural homes nestle in the grassy midlands of Swaziland. Today, both traditional circular huts and Western-style homes can be found in the country.

Swaziland is a small, beautiful country in southern Africa. It is surrounded by South Africa on three sides and by Mozambique on the east.

A Swazi village leader poses with his wives and son. Swazi men may have several wives, but each must have her own living quarters and garden.

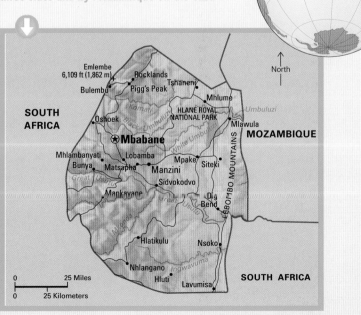

SWAZILAND

Sweden is a prosperous nation in northern Europe, a land of beautiful lakes, snow-capped mountains, swift rivers, and rocky offshore islands. Occupying more than half of the Scandinavian Peninsula it shares with Norway, Sweden is one of the largest European countries in area, and also one of the most thinly populated.

Sweden, Norway, and Denmark are called the Scandinavian countries. They share close cultural, economic, historical, language, and religious ties.

The people of modern Sweden enjoy one of the highest standards of living in the world. This is due largely to their skillful use of the country's natural resources of timber, iron ore, and water power. Taxes are high in Sweden, but the government provides many welfare benefits. Such benefits include free education, generous pensions, health benefits, and financial aid for housing.

Sweden's extensive welfare system provides every individual with a basic level of social and financial security. As a result, Sweden is considered by many people to be the model of an ideal society. Critics say the system makes people so secure that they get bored, but most Swedes support it.

Early settlement

Sweden's first inhabitants arrived in the southern tip of the country about 8000 B.C. About 50 B.C., the people in the region began trading with the Roman Empire, exchanging furs and amber for glass and bronze objects and silver coins.

Beginning about A.D. 800, Scandinavian pirates and warriors called *Vikings* sailed to many parts of the world, acquiring wealth by trade and conquest. Danish and Norwegian Vikings sailed mostly westward across the North Atlantic Ocean, but many Swedish Vikings sailed along rivers in eastern Europe. These Vikings eventually reached as far as the Black and Caspian seas. Their expeditions lasted until the 1000's.

Christianity became firmly established in Sweden during the reign of Olof Skotkonung, who ruled from the late 900's to the early 1000's. By that time, Sweden, Denmark, and Norway had become separate kingdoms.

During the 1200's and 1300's, there was a constant struggle for power between the nobles and the rulers of Sweden. The struggle drove the nobles to

SWEDEN

seek help from Queen Margaret of Denmark and Norway. In 1397, Sweden, Denmark, and Norway were brought together in a union that lasted more than 100 years.

seek help from Queen Margaret of Denmark and Norway. In 1397, Sweden, Denmark, and Norway were brought together in a union that lasted more than 100 years.

Independence and modern Sweden

Sweden broke away from the union in 1523. That year, Gustavus Vasa, a Swedish noble, defeated the Danes and became Gustav I Vasa, the first king of independent Sweden. Gustav laid the foundations of modern Sweden. He centralized the government, developed trade and industry, modernized the armed forces, and encouraged the Lutheran form of Christianity among his people. Gustav increased royal power but also expanded parliament to represent more of the people.

In 1630, King Gustav II Adolf, also known as Gustavus Adolphus, led an army into Germany to save the Protestants there from Roman Catholic domination. The king won several decisive victories, but he was killed on the battlefield in 1632.

Over the next two centuries, Sweden conquered many regions in northern Europe. However, its army was crushed by Peter the Great of Russia in the Battle of Poltava in 1709. Over the next few years, Sweden lost most of its European possessions.

The 1700's brought the Age of Liberty to Sweden. During this time, the Swedish *Riksdag* (parliament) passed a new constitution that transferred the king's power to the Riksdag. However, parliamentary rule ended in 1772, when an unsuccessful war in Germany and financial and political troubles at home led to a peaceful revolution that returned the king to power.

In the early 1800's, Sweden became involved in wars against the French Emperor Napoleon I. As a result of these wars, Sweden gained Norway from Denmark but lost Finland—a territory it had conquered in the 1100's and 1200's—to Russia. After the Napoleonic Wars ended in 1815, Sweden ended its involvement in military affairs.

In 1818, Jean Baptiste Jules Bernadotte, a French soldier who had become *regent* (acting ruler) of Sweden during the Napoleonic Wars, was elected king of Sweden as Charles XIV. Sweden's present royal family is descended from him.

243

SWEDEN TODAY

During the 1800's—before Sweden began to develop its manufacturing, mining, and forest industries—there was not enough work for the people, and food was often in short supply. By 1900, however, Sweden had begun to harness the power of its natural resources and develop into an important industrial nation. The nation's economy grew rapidly after World War II (1939–1945), and today Sweden ranks as one of the world's most prosperous nations.

In developing its economy and social system, Sweden has chosen what is often called the "middle way." Under this system, the government encourages private enterprise. At the same time, it seeks to direct business profits toward the creation of a society where every citizen is guaranteed a minimum income, free education, and largely free medical care. The government also pays pensions to retired people, widows, and orphans.

Since the 1990's, Sweden has begun to offer optional private health care. It has also cut back some social benefits, such as the amount of sick leave and unemployment compensation. Sweden made these changes because of an aging population and some slowing of economic growth had begun to strain the system.

On Jan. 1, 1995, Sweden became a member of the European Union (EU). In 2003, however, the people of Sweden rejected a proposal to replace their nation's currency, the *krona*, with the *euro*, the currency used by most EU nations.

Most of Sweden's people live in cities, and most cities lie mainly in the central and southern parts of the country. Many people in rural areas work part-time in farms and part-time in factories. The government also provides various types of support to farmers, thereby re-

FACTS

Official name:	Konungariket Sverige (Kingdom of Sweden)
Capital:	Stockholm
Terrain:	Mostly flat or gently rolling lowlands; mountains in west
Area:	173,732 mi² (449,964 km²)
Climate:	Temperate in south with cold, cloudy winters and cool, partly cloudy summers; subarctic in north
Main rivers:	Dal, Lule, Torne, Ume, Vindel
Highest elevation:	Mount Kebnekaise, 6,926 ft (2,111 m)
Lowest elevation:	Baltic Sea, sea level
Form of government:	Constitutional monarchy
Head of state:	Monarch
Head of government:	Prime minister
Administrative areas:	21 lan (counties)
Legislature:	Riksdag (Parliament) with 349 members serving four-year terms
Court system:	Hogsta Domstolen (Supreme Court)
Armed forces:	16,900 troops
National holidays:	Swedish Flag Day - June 6 (1916) National Day - June 6 (1983)
Estimated 2010 population:	9,243,000
Population density:	53 persons per mi² (21 per km²)
Population distribution:	84% urban, 16% rural
Life expectancy in years:	Male, 79; female, 83
Doctors per 1,000 people:	3.3
Birth rate per 1,000:	12
Death rate per 1,000:	10
Infant mortality:	3 deaths per 1,000 live births
Age structure:	0-14: 17%; 15-64: 65%; 65 and over: 18%
Internet users per 100 people:	80
Internet code:	.se
Languages spoken:	Swedish, Sami, Finnish
Religions:	Lutheran 75%, other Protestant 5%, Muslim 5%, other 15%
Currency:	Swedish krona
Gross domestic product (GDP) in 2008:	$484.55 billion U.S.
Real annual growth rate (2008):	0.7%
GDP per capita (2008):	$52,789 U.S.
Goods exported:	Chemicals, machinery, motor vehicles, paper products, wood products
Goods imported:	Chemicals, food, iron and steel, machinery, motor vehicles, petroleum products
Trading partners:	Denmark, Finland, France, Germany, Netherlands, Norway, United Kingdom, United States

ducing economic and social differences between urban and rural life. Most Swedish homes are small, with only a few rooms, but many families also have a small vacation home.

All children between the ages of 7 and 16 must attend school. Students are required to study English from the fourth through the seventh grade, and most continue English after that.

The Swedes are active people who enjoy the outdoors. Cross-country skiing and hockey are the chief winter sports. Bicycling, camping, hiking, sailing, soccer, swimming, and tennis are also popular.

Most Swedes are descendants of ancient Germanic tribes who settled in Scandinavia beginning around 8000 to 5000 B.C. People of Finnish origin make up the country's largest ethnic minority. Sami, also known as Lapps, live in the northernmost part of the country. Since the late 1900's, many people from other countries have migrated to Sweden to find jobs or to escape from areas of conflict around the world.

One of the largest countries in Europe, Sweden is also one of the most thinly populated. Most Swedes live in the center and south of the country. About one-third of the population lives in the three largest cities—Stockholm, Göteborg, and Malmö.

Riddarholmen (the Isle of Knights) is one of the 14 islands that make up Stockholm, Sweden's capital and largest city. One of the oldest parts of the city, the island features the Riddarholm Church, which dates from the late 1200's.

FIGHTING FOR PEACE

Throughout much of its early history, Sweden was a warlike nation. Beginning in the 800's, Swedish Vikings raided settlements in the Baltic and along the rivers of eastern Europe. In 1630, the Swedish king Gustavus Adolphus led 13,000 troops into battle during the Thirty Years' War, the last of Europe's great religious wars. Two centuries later, Swedish soldiers battled French forces in the Napoleonic Wars.

Since the Napoleonic Wars, however, Sweden has stopped taking an active part in military affairs, and the nation remained neutral during World War I (1914-1918) and World War II (1939-1945). Through the years, Sweden has worked to build peace between nations, strongly supporting organizations such as the League of Nations and its successor, the United Nations (UN).

Wartime heroes

Despite Sweden's neutral position in World War II, many Swedes worked and fought—and sometimes died—to save others during the bloody conflict.

One such hero was Raoul Wallenberg, a businessman and diplomat who risked his life to help save about 100,000 Hungarian Jews from being killed by the Nazis in 1944. In 1945, Soviet forces captured and imprisoned Wallenberg as a suspected spy. In 1957, the Soviet government announced that Wallenberg had died of a heart attack in 1947. However, several people reported seeing him in Soviet prisons and hospitals after 1947.

During the 1950's, Dag Hammarskjöld, a Swedish statesman, worked to ease tension between the United States and the Soviet Union. In 1955, he secured the release from China of American prisoners captured during the Korean War (1950-1953). In 1956, Hammarskjöld helped solve the Suez crisis between Egypt and Israel and Israel's allies—France and the United Kingdom.

Other Swedes who have worked for peace between nations include Alva R. Myrdal, who became famous for promoting nuclear disarmament. From 1962 to 1973, she headed Sweden's delegation to

THE NOBEL PRIZES

The king of Sweden attends the annual Nobel Prize ceremony.

A bas relief likeness of Alfred Nobel appears on the front side of each Nobel Prize medal.

Albert Schweitzer who received the Nobel Peace Prize in 1952, was a brilliant philosopher, musician, physician, and missionary. He devoted most of his life to humanitarian work in Africa.

Dag Hammarskjöld served as secretary-general of the United Nations from 1953 to 1961. He was awarded the 1961 Nobel Peace Prize.

Al Gore won the 2007 Nobel Peace Prize for his work in spreading awareness of climate change.

Woodrow Wilson
(Nobel Prize 1919)

SOME NOBEL PRIZE WINNERS

Nobel Prizes for Physics

1901 Wilhelm C. Roentgen (German) for discovering X rays.

1921 Albert Einstein (German) for contributing to mathematical physics and stating the law of the photoelectric effect.

1945 Wolfgang Pauli (Austrian) for discovering the exclusion principle (Pauli principle) of electrons.

Nobel Prizes for Chemistry

1911 Marie Curie (French) for her discovery of radium and polonium, and for her work in isolating radium and studying the compounds of radium.

1960 Willard F. Libby (American) for developing a method of radiocarbon dating.

1964 Dorothy C. Hodgkin (British) for X-ray studies of compounds such as vitamin B12 and penicillin.

Nobel Prizes for Physiology or Medicine

1924 Willem Einthoven (Dutch) for discovering how electrocardiography works.

1930 Karl Landsteiner (American) for discovering the four main human blood types.

2005 Barry J. Marshall and J. Robin Warren (Australian) for discovering the bacterium Helicobactor pylori and its role in ulcer disease.

Nobel Prizes for Literature

1907 Rudyard Kipling (British) for his stories, novels, and poems.

1925 George Bernard Shaw (Irish-born) for his plays.

1970 Alexander Solzhenitsyn (Russian) for his novels.

1993 Toni Morrison (American) for her novels.

2005 Harold Pinter (British) for his plays.

Nobel Prizes for Peace (Awarded in Norway)

1901 Jean Henri Dunant (Swiss) for founding the Red Cross and originating the Geneva Convention, and Frédéric Passy (French) for founding a French peace society.

1919 Woodrow Wilson (American) for seeking a just settlement of World War I.

1961 Dag Hammarskjöld is awarded the prize posthumously (after his death), for his work as secretary general of the United Nations.

1964 Martin Luther King, Jr. (American), for leading the African American struggle for equality in the United States.

1971 Willy Brandt (German) established policies to improve West German relations with East Germany, Poland, and the Soviet Union.

1979 Mother Teresa (Indian) for aiding India's poor.

2011 President Ellen Johnson-Sirleaf (Liberian), peace activist Leymah Gbowee (Liberian), and prodemocracy activist Tawakkul Karman (Yemeni) for their work toward the safety of women and women's rights.

Willy Brandt
(Nobel Prize 1971)

Mother Teresa
(Nobel Prize 1979)

Nelson Mandela
(Nobel Prize 1993)

the United Nations Disarmament Conference in Geneva, Switzerland. Myrdal shared the 1982 Nobel Peace Prize with Alfonso García Robles of Mexico for their contributions to UN disarmament negotiations.

The Nobel Prizes

The Nobel Prizes were established by Alfred Nobel, the Swedish chemist and industrialist who invented dynamite. After Nobel received a patent for his invention in 1867, the sale of dynamite brought him great wealth. In his will, he set up a fund of about $9 million and directed that the interest from the money was to be used to award five annual prizes to people whose work most benefited humanity.

First presented in 1901, the Nobel Prizes are given for the most important discoveries or inventions in the fields of physics, chemistry, and physiology or medicine; the most distinguished literary work of an idealistic nature; and the most effective work in the interest of international peace. A sixth prize—the Bank of Sweden Prize in Economic Sciences in Memory of Alfred Nobel—was established by the Bank of Sweden and first awarded in 1969.

The Royal Academy of Sciences in Stockholm chooses the physics, chemistry, and economics winners. The Nobel Assembly at the Karolinska Institute in Stockholm awards the prize for medicine. The Swedish Academy in Stockholm awards the prize for literature. The prize for peace is the only Nobel Prize that is not administered in Sweden. It is awarded by the Norwegian Nobel Committee elected by the Norwegian *Storting* (parliament).

Today, Nobel's legacy lives on in the hearts of the Swedish people. Sweden continues to be a strong supporter of international peace and disarmament. The nation's long-standing concern for the disadvantaged is evident in its actions in world affairs. Sweden provides sanctuary for refugees, and it is a major supplier of aid to many countries.

ENVIRONMENT

The spectacular beauty of Sweden's landscape makes it a paradise for nature lovers. The country's scenic features range from the snow-capped mountains of the far northern region to the fertile green plains of the southern lowlands. Thousands of lakes cover much of the country's area, and many small groups of islands lie off Sweden's coast. Miles of sandy beaches line the southern coast, and rocky cliffs rise steeply from the sea in the north.

Sweden occupies the eastern half of the Scandinavian Peninsula. It is bordered to the west by Norway, to the north and northeast by Finland, and to the east by the Gulf of Bothnia and the Baltic Sea. Denmark lies off the southwest coast. The northernmost part of Sweden is north of the Arctic Circle.

Sweden's climate varies greatly between the Arctic north and the temperate south. Southwesterly winds from the Atlantic Ocean bring pleasant summers and generally mild winters to southern Sweden. Summers are also pleasant in northern Sweden, but winters are cold. The Kölen Mountains prevent the warming winds from reaching the northern regions.

Sweden's abundant wildlife includes hares, foxes, red squirrels, weasels, moose, lynx, and elk. Red deer are common in the southern regions. Ducks and swans nest in the lake regions, while such birds as owls, ptarmigans, grouse, and woodcock inhabit the forests. The country's swift-flowing rivers teem with salmon, trout, and pike.

Northern land regions

Sweden has four main land regions: the Mountain Range, the Inner Northland, the Swedish Lowland, and the South Swedish Highland.

The Mountain Range is part of the Kölen Mountains. Sweden's northern boundary with Norway runs through these mountains, which rise to about 6,600 feet (2,000 meters). Hundreds of small glaciers cover the higher slopes of this snow-capped range. Much of the Mountain Range region lies north of the Arctic Circle, where the bitter cold above 1,600 feet (488 meters) prevents tree growth. However, some birch trees grow on the southern slopes.

Lovely Lake Siljan lies in central Sweden. Church boats that look like Viking vessels once carried villagers to lakeside churches on Sollerön, an island in the lake.

The granite rocks of Bohuslän have been worn smooth over thousands of years by the action of wind, water, and ice. Bohuslän, the most westerly of Sweden's provinces, extends from Göteborg north to the Norwegian border.

The craggy shoreline of Gotland is lined with limestone pillars known as raukar. Gotland, the largest of Sweden's islands, lies off the country's southeast coast in the Baltic Sea. During the 1100's, the island was an important European trading center.

Birch forests provide a beautiful setting for a stuga (country cottage) near Uppsala, northwest of Stockholm. Sweden's main timber regions are located in the north and north-central regions, where birch, pine, and spruce trees thrive.

East and south of the Mountain Range lies the Inner Northland—a vast, thinly populated, hilly region that slopes gently eastward toward the Gulf of Bothnia. The great forests of pine and spruce trees that cover most of the land are crossed by swift rivers and their deep valleys, some with long lakes.

The evergreen forests give way to deciduous hardwood trees in the more southerly district of Dalarna, around Lake Siljan. Some of Sweden's most colorful midsummer celebrations take place in this region, earning it the nickname of the *Folklore District*.

Southern Sweden

The Swedish Lowland, the southern third of the country, has more people than any other part of the country. This region is made up of broad central plains broken by lakes, tree-covered ridges, and small hills. Many lakes in this region—including Vänern and Vättern, Sweden's two largest lakes—were cut by glaciers and filled by melting ice after the last Ice Age ended.

Another lake in this region, Mälaren, had outlets that opened to the sea until as recently—in geologic terms—as the A.D. 1200's. Mälaren was finally enclosed by land that had been rising slowly since the end of the last Ice Age. The Göta Canal links Sweden's east and west coasts by connecting several lakes and rivers. It was one of the engineering wonders of the world when it was built in the early 1800's.

The South Swedish Highland, also known as the Götaland Plateau, is a rocky upland rising to about 1,200 feet (366 meters) at the southern tip of the country. The southern part of the region is flat, with small lakes and swamps. This sparsely populated area has poor, stony soils, covered mainly by forests.

Farmland covers more than 40 percent of the Swedish Lowland's fertile, rolling plains. Skåne, in the far south, is the richest farming area. The Swedish Lowland also includes many lakes, as well as major cities.

Sweden's land area includes a long coastline with many offshore islands. Treeless tundra and dense forests dominate the cold northern regions, which are inhabited by the Sami. Most Swedes live in the warmer, southern regions.

STOCKHOLM

Sergels Torg, with its towering obelisk and magnificent fountains, marks the center of modern Stockholm. Its striking, modern architecture contrasts sharply with the buildings of the city's older sections.

The narrow lanes and quaint architecture of Gamla Stan, Stockholm's historic quarter, reflect Stockholm's origins as a shipping and trading city during the Middle Ages. Sweden's Riksbank, the oldest existing central bank in the world, was founded in 1656.

Stockholm, Sweden's capital and largest city, stands in a magnificent natural setting among heavily wooded hills. Stockholm is the heart of the nation's commercial and cultural life and also serves as a major center for international trade and communications. The University of Stockholm; Sweden's Royal Ballet, Library, Opera, and Dramatic Theater; and numerous city museums and galleries reflect the city's cultural splendor. One of the most beautiful, spacious cities in the world, Stockholm is filled with parks, tree-lined squares and boulevards, playgrounds, and wading pools.

Stockholm is built on 14 islands and a part of the mainland. It is connected by about 50 bridges. The city lies on the east coast of Sweden, between Lake Mälaren and the Baltic Sea. Thousands of islands in the sea east of Stockholm form an archipelago.

Stockholm has a pleasant, relaxed atmosphere, and the contrasts of land and water and of old and new add to its unique charm. Careful ity planning has ensured that fine old buildings dating from the Middle Ages blend with striking, modern structures in a pleasing way.

Stockholm's long tradition of sensible city planning has helped the city overcome such problems as congestion and urban decay. For example, to cope with the rapidly growing population in the years after World War II, city officials built residential suburbs on land that the government had purchased as long ago as 1904.

The Old Town

At the heart of Stockholm, on the island of Stadsholmen, lies the *Gamla Stan* (Old Town), where the city began. In the 1250's, a Swedish nobleman called Birger Jarl built a castle in the area that is now Gamla Stan. At about the same time, Stockholm was granted a charter, giving it the status of a town and assuring it royal protection.

A German influence is evident in the streets of Gamla Stan, which follow the

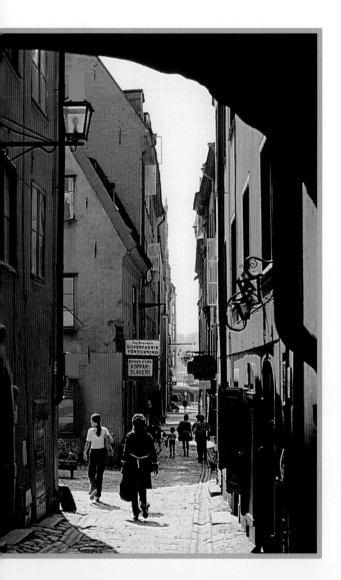

routes established during medieval times. The *Tyska kyrkan* (German Church), for example, stands on the Skomakaregatan.

The Royal Palace can be seen on the island of Stadsholmen in the main part of Gamla Stan. The palace was built in the early 1700's on the same site as the original palace, which burned to the ground in 1697. Its historic and artistic treasures include the monarch's silver throne and pews saved from the original palace.

An expanding city

As Swedish influence spread during the 1600's, Stockholm grew into a prosperous urban center. Norrmalm, once a community north of Gamla Stan, is now the commercial center of Stockholm, and its busy streets are lined with office buildings and shops. Much of the area was rebuilt in the 1950's and 1960's, with sections restricted for pedestrian traffic—a pioneering move in city planning at that time.

Stockholm continued to expand into the late 1600's to cover the large island of Södermalm, which can be reached by crossing a bridge from Gamla Stan. Between Södermalm and Östermalm lies the large island of Djurgården, which features an enormous park and open-air museum known as Skansen. Farmhouses, windmills, and barns brought to Skansen from every part of the country give visitors a taste of traditional rural Sweden.

Another Djurgården attraction is the *Vasa* museum. The battleship *Vasa* was the most powerful fighting ship of its day in the Baltic. Launched in 1628, the ship sank on its maiden voyage and was not rediscovered until 1956. The *Vasa* was raised in 1961 and then painstakingly restored.

Strandvägen, one of Stockholm's finest streets, can be seen across the inlet of Nybrovik. The islands in the sea east of Stockholm are popular vacation spots for Swedish citizens, who also enjoy boating in the city's waterways.

ECONOMY

Sweden is a highly industrialized nation. Its economy is based on a combination of advanced engineering and service industries. It also relies heavily on exports. Most Swedish industry is privately owned. Government ownership is concentrated in mines, public transportation, energy, and telecommunications.

Between the mid-1800's and mid-1900's, Sweden's natural resources helped change the country from a poor agricultural country to an advanced industrial nation. Close cooperation among the government, employer groups, and labor unions also has helped bring about Sweden's economic development.

Iron and steel

Sweden has some of the richest and highest-grade iron ore deposits in the world. Most of the iron ore is mined near Kiruna in Lapland. Workers use high-quality steel made from the iron ore to manufacture ball bearings, stainless steel goods, and precision tools.

Steel is widely used in the production of such engineering products as agricultural machinery, aircraft, automobiles, and ships. The engineering industry accounts for about half of Sweden's industrial production and about half of its exports. The electrical engineering industry makes equipment for power plants and communications. Telephones are an important export.

The Swedish chemical industry imports most of its raw materials. The chief products include explosives, pharmaceuticals, plastics, and safety matches. Safety matches were invented in Sweden in 1844, and the country is still one of the world's leading producers.

Agriculture, forestry, and fishing

Although only about 10 percent of Sweden's land is suitable for agriculture, Swedish farmers produce almost enough food to feed all the people. The best farmland lies in central and southern

Economic uses of land in Sweden and the country's main farm products and chief mineral deposits are shown above. Major manufacturing centers are in red.

The old iron ore town of Kiruna boasts a modern town hall. The town has become a center of scientific research.

Lumber floats down the Klar River to the old woodworking town of Karlstad, on Sweden's western coast.

Rista Falls in central Sweden provides a scenic backdrop for kayakers. In addition to offering a white-water challenge for adventurous tourists, Sweden's swift-flowing rivers provide the country with valuable hydroelectric power.

Sweden, especially in Skåne. Most Swedish farmers own their own land. Almost all farmers belong to a *cooperative,* which collects, processes, and markets their farm products.

Dairy farming and livestock raising are important sources of income for Swedish farmers. Chickens are raised for eggs and meat. The main crops are barley, oats, potatoes, sugar beets, and wheat.

Large quantities of fish are caught in Sweden's coastal waters. The most important saltwater varieties are cod, herring, mackerel, prawns, and salmon.

Forests cover more than half of Sweden, and much of the nation's exports are timber or products made of wood. The evergreen forests of the north provide most of the commercially valuable timber. Forestry is less important in southern Sweden. Although much lumber is carried to manufacturing plants by truck and railroad, logs are also floated down the rivers to sawmills on the coast.

Water power

Sweden's rivers, waterfalls, and lakes provide enough water power to supply much of the country's electricity. The peak production periods of the hydroelectric power stations in northern and southern Sweden occur in different seasons, so the stations are coordinated to supply adequate water power at all times.

Service industries

Service industries provide about three-fourths of Sweden's jobs and produce over two-thirds of the total value of Sweden's economic production. Service industries provide services rather than produce goods.

Community, government, and personal services are by far the leading employers among the service industries in Sweden. Among the most important fields are education, health care, information and communication technology, and scientific research and development.

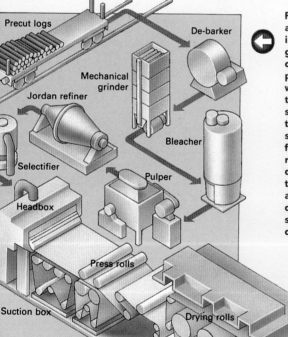

Paper production is an important industry in Sweden. The diagram shows how precut logs are processed into pulp, which moves through the selectifier (a pressurized sieve) into the headbox. A high-speed belt takes the fibers across water-removing suction devices, and then through press rolls and drying rolls. The dried paper is smoothed out in the calender rolls.

Precut logs
De-barker
Mechanical grinder
Jordan refiner
Bleacher
Selectifier
Pulper
Headbox
Press rolls
Suction box
Drying rolls
Calender rolls

SWITZERLAND

Switzerland is a small European country known for its beautiful, snow-capped mountains, including the majestic Matterhorn in the Alps. The Alps and the Jura Mountains cover more than half of the country, but most of the Swiss people live on a plateau that lies between the two mountain ranges. The plateau region contains most of Switzerland's industries and its richest farmlands. Bern, the country's capital city, and Zurich, its largest city, are also located there.

The Swiss have a tradition of independence that dates back about 700 years. At that time, people in what is now central Switzerland agreed to help one another resist foreign rule. Gradually, people in nearby areas joined them and formed the Swiss Confederation. But although these groups joined forces, each kept its own way of life. As a result, various Swiss groups have different languages, customs, and traditions.

The freedom-loving Swiss are united in a firm decision to avoid involvement in wars. In the early 1500's, Switzerland established a policy of neutrality that continues today. During World War I (1914-1918) and World War II (1939-1945), the country provided a safe haven for thousands who fled from the fighting and from political persecution. Switzerland has even been reluctant to join international organizations that might bring its neutrality into question. Thus, although Geneva, Switzerland, had been the European headquarters of the United Nations (UN) since 1946, Switzerland did not join the organization until 2002.

The country's policy of neutrality has helped develop its international banking services. Banking is one of Switzerland's major industries. Swiss banks are safer than those in most other countries, and so they attract customers from all over the world.

Switzerland is also famous for its high-quality manufactured products. Swiss watches, for instance, are highly valued throughout the world. Other famous Swiss products include electrical equipment, industrial machinery, precision instruments, chocolate, and cheese.

A number of celebrated artists and writers have come from Switzerland. Outstanding Swiss artists have included painter Paul Klee and sculptor Alberto Giacometti. Another Swiss, Le Corbusier, won fame in modern architecture. Notable Swiss writers have included Friedrich Dürrenmatt, Max Frisch, and Charles Ferdinand Ramuz. Two children's classics were also written by Swiss authors—*Heidi* (1881) by Johanna Spyri and *The Swiss Family Robinson* (1812-1813) by the Wyss family.

SWITZERLAND TODAY

Switzerland is a confederation that functions as a federal republic. Its government is based on the Swiss Constitution of 1848, which was changed in 1874 and 1999. The Constitution divides political powers between the central government and *cantonal,* or state, governments. There are 20 cantons and 6 half-cantons. The country's half-cantons were originally undivided cantons, but they split into separate political units with as much power of self-government as the full cantons.

Switzerland's legislature is composed of a two-house Federal Assembly—the Council of States and the National Council. The Council of States is composed of 46 members. Each of the 20 cantons elects two members, and each of the 6 half-cantons elects one member. The 200 members of the National Council are elected to four-year terms from election districts based on population.

A seven-member cabinet called the Federal Council serves in place of a single chief executive. Cabinet members are elected by the legislature to four-year terms. The president, Switzerland's head of state, is chosen by the legislature from among the Federal Council members and serves a one-year term. The president's duties are largely ceremonial.

Swiss citizens enjoy close control over their laws. The right of *referendum* allows the people to demand a popular vote on laws passed by the legislature. The right of *initiative* gives the people the right to bring up specific issues for a vote by the people.

An example of Switzerland's form of direct democracy is the open-air meeting, or *Landsgemeinde,* which still is held in Canton Glarus and in the half-canton of Appenzell Innerrhoden. The people in these regions vote on important issues by a show of hands.

Land

Switzerland has three main land regions—the Jura Mountains, the Swiss Plateau, and the Swiss Alps. The Jura Mountains consist of a series of parallel ridges

FACTS

Official name:	Schweizerische Eidgenossenschaft, in German; Confédération Suisse, in French; Confederazione Svizzera, in Italian (all meaning Swiss Confederation)
Capital:	Bern
Terrain:	Mostly mountains (Alps in south, Jura in northwest) with a central plateau of rolling hills, plains, and large lakes
Area:	15,940 mi² (41,284 km²)
Climate:	Temperate, but varies with altitude; cold, cloudy, rainy/snowy winters; cool to warm, cloudy, humid summers with occasional showers
Main rivers:	Rhine, Rhône, Inn, Ticino
Highest elevation:	Dufourspitze, 15,203 ft (4,634 m)
Lowest elevation:	Lake Maggiore, 633 ft (193 m)
Form of government:	Confederation
Head of state:	President
Head of government:	President
Administrative areas:	26 kantone, in German; cantons, in French; cantoni, in Italian (all meaning cantons)
Legislature:	Federal Assembly consisting of the Council of States with 46 members serving four-year terms and the National Council with 200 members serving four-year terms
Court system:	Federal Supreme Court
Armed forces:	22,800 troops
National holiday:	Founding of the Swiss Confederation - August 1 (1291)
Estimated 2010 population:	7,595,000
Population density:	476 persons per mi² (184 per km²)
Population distribution:	74% urban, 26% rural
Life expectancy in years:	Male, 79; female, 84
Doctors per 1,000 people:	4.0
Birth rate per 1,000:	10
Death rate per 1,000:	8
Infant mortality:	4 deaths per 1,000 live births
Age structure:	0-14: 16%; 15-64: 68%; 65 and over: 16%
Internet users per 100 people:	76
Internet code:	.ch
Languages spoken:	German (official), French (official), Italian (official), Romansh
Religions:	Roman Catholic 41.8%, Protestant 35.3%, Muslim 4.3%, other 18.6%
Currency:	Swiss franc
Gross domestic product (GDP) in 2008:	$491.22 billion U.S.
Real annual growth rate (2008):	1.9%
GDP per capita (2008):	$65,131 U.S.
Goods exported:	Chemicals, machinery, pharmaceutical products, precision instruments, watches
Goods imported:	Agricultural products, chemicals, machinery, medicines, motor vehicles, petroleum products
Trading partners:	France, Germany, Italy, United States

Switzerland is bordered by France, Germany, Liechtenstein, Austria, and Italy. The Jura Mountains and the Swiss Alps make up about 65 percent of the country's total area. Most of the Swiss people live on the plateau that runs between these two mountain ranges.

separated by narrow valleys. The ridges extend along Switzerland's western border and into France.

The Swiss Plateau, a hilly region with rolling plains, has Switzerland's richest farmland, as well as Lake Constance and Lake Geneva. About 80 percent of the Swiss people live in this region.

The Swiss Alps—a heavily forested region covering about 60 percent of Switzerland— are part of the mighty Alps mountain system. The upper valleys of the Rhine and Rhône rivers divide the Swiss Alps into a northern and a southern series of ranges. The scenic landscape attracts millions of visitors.

Climate

The climate in each land region differs greatly because of the wide variety in altitude. The Swiss Plateau has warm summers and cool winters. Valleys in some southern cantons are warmer. Occasionally, a dry, warm, southerly wind called the *foehn* blows down the valleys of the Swiss Alps, causing an early melting of the mountain snows. Flooding or avalanches may then occur.

A traffic officer keeps watch on the busy streets of Bern, the Swiss capital. The center of the city lies in a loop of the river Aare. The city was founded in 1191 and joined the Swiss Confederation in 1353. It is now the capital of the canton (state) of Bern, the second most populous in Switzerland.

CANTONS

In 1273, Rudolf I of Switzerland's Habsburg family became Holy Roman Emperor, and some Swiss people began to fear the Habsburg family's growing power. The free men of what are now the cantons of Uri, Schwyz, and Unterwalden decided to band together to protect their freedom.

In August 1291, leaders of the three regions signed the Perpetual Covenant and promised to aid each other against foreign rulers. Their pledge was the start of the Swiss Confederation. The confederation came to be known as *Switzerland,* taking its name from the canton of Schwyz.

In the 1300's, the Swiss fought and won three wars of independence against the Habsburgs, who by then were based in Austria. Many famous stories were told about Swiss heroes who took part in these wars, including the legendary William Tell. According to legend, Tell started a Swiss revolt by shooting an arrow through the heart of a tyrannical Austrian bailiff.

During the 1400's, Switzerland became a strong military power and entered several wars to gain land. In 1515, however, the French dealt the Swiss a crushing defeat at Marignano in Italy. As a result, Switzerland soon abandoned its policy of expansion and adopted one of neutrality. The Swiss fought several internal wars during the 1500's to 1700's, and France conquered them in 1798 at the time of the French Revolution. After the Napoleonic Wars, Swiss neutrality was guaranteed at the Congress of Vienna (1814-1815).

By 1830, many Swiss had begun to demand political reforms and individual rights. Following a three-week civil war, Switzerland adopted a new constitution in 1848 that guaranteed the rights of its citizens.

The Constitution provided for a democratic central government. But it also allowed the cantons to retain their independence. Each of the country's cantons and half-cantons has its own constitution, legislature, and executive council. Nevertheless, the Swiss stand together as one people. As a symbol of their unity, during the national festival celebrated every August 1, the Perpetual Covenant is commemorated by lighting a chain of beacons that shines across the country.

Schaffhausen
Canton since 1501
Area: 115 sq mi (298 km²)
Population: 73,392
Capital: Schaffhausen
Majority language: German

Thurgau
Canton since 1803
Area: 383 sq mi (991 km²)
Population: 228,875
Capital: Frauenfeld
Majority language: German

Basel-Land 1501;
half-canton since 1833
Area: 200 sq mi (518 km²)
Population: 259,374
Capital: Liestal
Majority language: German

Basel-Stadt 1501;
half-canton since 1833
Area: 14 sq mi (37 km²)
Population: 188,079
Capital: Basel
Majority language: German

Aargau
Area: 542 sq mi (1,404 km²)
Population: 547,493
Capital: Aarau
Majority language: German

Zurich
Canton since 1351
Area: 668 sq mi (1,729 km²)
Population: 1,247,906
Capital: Zurich
Majority language: German

St. Gallen
Canton since 1803
Area: 782 sq mi (2,026 km²)
Population: 452,837
Capital: St. Gallen
Majority language: German

Appenzell Ausserrhoden 1513;
half-canton since 1597
Area: 94 sq mi (243 km²)
Population: 53,504
Capital: Herisau
Majority language: German

Appenzell Innerrhoden 1513;
half-canton since 1597
Area: 66 sq mi (173 km²)
Population: 14,618
Capital: Appenzell
Majority language: German

Jura
Canton since 1979 (previously part of Bern)
Area: 324 sq mi (839 km²)
Population: 68,224
Capital: Delémont
Majority language: French

Solothurn
Canton since 1481
Area: 305 sq mi (791 km²)
Population: 244,341
Capital: Solothurn
Majority language: German

Lucerne
Canton since 1332
Area: 576 sq mi (1,493 km²)
Population: 350,504
Capital: Lucerne
Majority language: German

Zug
Canton since 1352
Area: 92 sq mi (239 km²)
Population: 100,052
Capital: Zug
Majority language: German

The annual Landsgemeinde meeting takes place in Appenzell, the capital of the half-canton Appenzell Innerrhoden.

Switzerland is divided into 26 government units: 20 cantons and 6 half-cantons. The half-cantons were originally three undivided cantons. They split into separate political units with as much power of self-government as the full cantons.

Schwyz
Canton since 1291
Area: 351 sq mi (908 km²)
Population: 111,964
Capital: Schwyz
Majority language: German

Glarus
Canton since 1352
Area: 264 sq mi (685 km²)
Population: 38,183
Capital: Glarus
Majority language: German

Graubünden
Canton since 1803
Area: 2,744 sq mi (7,106 km²)
Population: 187,058
Capital: Chur
Majority language: German

Uri
Canton since 1291
Area: 416 sq mi (1,077 km²)
Population: 34,777
Capital: Altdorf
Majority language: German

Unterwalden Nidwalden 1291;
half-canton since 1340
Area: 107 sq mi (276 km²)
Population: 37,235
Capital: Stans
Majority language: German

Unterwalden Obwalden 1291;
half-canton since 1340
Area: 189 sq mi (491 km²)
Population: 32,427
Capital: Sarnen
Majority language: German

Bern
Canton since 1353
Area: 2,301 sq mi (5,959 km²)
Population: 957,197
Capital: Bern
Majority language: German

Neuchâtel
Canton since 1815
Area: 310 sq mi 803 km²)
Population: 167,949
Capital: Neuchâtel
Majority language: French

Vaud
Canton since 1803
Area: 1,240 sq mi (3,212 km²)
Population: 640,657
Capital: Lausanne
Majority language: French

Fribourg
Canton since 1481
Area: 645 sq mi (1,671 km²)
Population: 241,706
Capital: Fribourg
Majority language: French

Valais
Canton since 1815
Area: 2,017 sq mi (5,224 km²)
Population: 272,399
Capital: Sion
Majority language: French

Ticino
Canton since 1803
Area: 1,086 sq mi (2,811 km²)
Population: 306,846
Capital: Bellinzona
Majority language: Italian

Geneva
Canton since 1815
Area: 109 sq mi (282 km²)
Population: 413,673
Capital: Geneva
Majority language: French

PEOPLE

Switzerland is famous for its spirit of independence, and each region maintains its own language, customs, and traditions. The Swiss Constitution, which provides for four national languages, including three official languages, upholds the people's desire for independence.

Languages

The official languages are German, French, and Italian, and national laws are published in all three languages. The four national languages include the three official ones plus *Romansh*, which is closely related to Latin. Romansh is spoken by people in the mountain valleys of the southeastern canton of Graubünden.

About two-thirds of the people speak a form of German called *Schwyzerdütsch*. Yet even among this German-speaking majority—who live in the northern, eastern, and central parts of Switzerland—dialects vary from place to place. Schwyzerdütsch and its variants are used in conversation, but standard German is used in newspapers, books, television and radio programs, plays, and church sermons.

French is spoken in western Switzerland by about 20 percent of the people. And Italian is used by nearly 7 percent of the population, in the south. French and Italian, as spoken by the Swiss, vary little from their standard forms.

The linguistic variety in Switzerland has caused little serious conflict. In the 1970's the French-speaking minority in the canton of Bern campaigned to separate from the largely German-speaking region and form a canton of their own. This resulted in the creation of the canton of Jura in 1979.

People in the cantons celebrate their own traditions in local festivals that include singing, band music, and folk dancing in colorful national costumes.

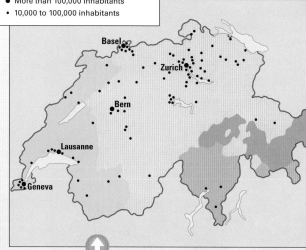

The Lavaux vineyards in French-speaking western Switzerland produce fine white wine. Like their neighbors in France to the west, the French-speaking Swiss are master winemakers.

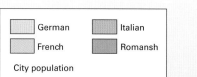

	German		Italian
	French		Romansh

City population
- More than 100,000 inhabitants
- 10,000 to 100,000 inhabitants

Basel · Zurich · Bern · Lausanne · Geneva

The different languages spoken in Switzerland reflect the diverse origins of its people. About 70 percent speak German, about 20 percent speak French, about 7 percent speak Italian, and less than 1 percent speak Romansh.

A man in Zurich walks past a display at a Swiss bank showing current stock prices. Zurich is one of the world's financial centers.

UBS

Population and recreation

Switzerland has a total population of more than 7-1/2 million people, and no Swiss city has a population of more than 500,000. Bern is the country's capital, and Zurich is the largest city. Other large cities include Basel, Geneva, and Lausanne. About one-fourth of the Swiss people live in rural areas.

For centuries, Switzerland has welcomed people from other countries who sought relief from political and religious oppression. Because Switzerland has more jobs than its own people can fill, many foreigners work in Switzerland. About one-fifth of the nation's inhabitants are non-Swiss citizens. The largest groups come from Germany, Italy, Portugal, Spain, Turkey, and the nations that were part of the former Yugoslavia. Many political refugees also have found asylum in Switzerland.

The mountains of Switzerland provide abundant recreational opportunities for all the Swiss people. Many of the people enjoy skiing, bobsledding, camping, climbing, hiking, bicycling, and boating. Gymnastics, target shooting, and soccer are also popular. *Hornussen* is a very old game that is somewhat like baseball. Two teams play Hornussen. A batter uses a long wooden club to try to hit a wooden disk, and fielders use wooden rackets to catch the disk.

An Italian-speaking Swiss man shares a traditional love of good food. Located in southern Switzerland, the Italian-speaking regions enjoy the country's warmest weather.

Many of the German-speaking Swiss in the Alpine region are descendants of people who had settled in Switzerland by the A.D. 400's. The German-speaking majority lives in the northern, eastern and central parts of the country.

ECONOMY

Switzerland has one of the most prosperous economies in the world. The nation's highly specialized industries are extremely profitable, and there is little unemployment.

Because the country lacks the important mineral resources on which heavy industry is based, Switzerland depends heavily on imported raw materials. Its industries process these materials into high-quality products for export. Income from banking, tourism, and other services also contribute greatly to Switzerland's economic strength.

Industry and agriculture

To keep down the costs of the imported raw materials, Switzerland's industries specialize in skilled precision work on small, valuable items. For example, Switzerland is famous for producing high-quality watches. Most Swiss factories are small- or medium-sized because of the stress on quality goods rather than mass production.

More than 95 percent of all watches manufactured in Switzerland are exported. Another important export industry is the manufacture of mechanical engineering products. These products include generators and other electrical equipment, industrial machinery, precision instruments, and motor vehicles. Other major Swiss products are pharmaceuticals; chemicals; paper; processed foods, including cheese and chocolate; and silk and other textiles.

Most of Switzerland's industries depend on hydroelectricity, which is generated at power stations on the country's rapidly flowing rivers. The use of hydroelectricity to power factories and railroads helps keep the busiest industrial centers almost smoke-free. Several nuclear power plants supply most of the rest of the country's electric power.

Switzerland's transportation system is excellent, despite the difficulties of travel through the mountains. The government owns and operates most of the railroad network. Paved roads and highways provide travel even to mountain areas, although the

roads through some high mountain passes are closed due to snow for certain parts of the year.

Farmers can raise crops on only about 10 percent of Switzerland's total area, but modern farming methods make the land as productive as possible. Crops include fruits, grains, and potatoes. Grapes and olive trees grow in sunny areas in western and southern Switzerland. Agriculture supplies only about half of the people's needs; the rest of the nation's food is imported.

Livestock provides most of Switzerland's farm income, largely through dairy farming. In the summer, most of the dairy cattle graze on the high mountain pastures that cover about one-fourth of the country. In the winter, the cattle are brought down to the valleys. Much of the milk is used to make such cheeses as Emmentaler (also known as Swiss cheese) and Gruyère for export. Farmers also raise hogs, goats, sheep, and chickens.

Saint Moritz is a famous resort town in the Alps of eastern Switzerland. The town stands between mountain slopes and along a small lake. Favorite activities there include skiing, ice skating, sailing, swimming, and hiking. Tourism is a major contributor to the Swiss economy.

Pharmaceuticals and chemicals are among Switzerland's major industrial products. Several important chemical companies are located in Basel in the northern part of the country.

Careful packing is the final stage in the production of Swiss chocolate, a favorite around the world.

Strict controls ensure consistent quality of Emmentaler cheese, or Swiss cheese. The cheese is one of the country's best-known products. Cheese ranks among Switzerland's principal food exports.

Banking and tourism

People from many foreign countries invest in Switzerland's banks, which are located mainly in Zurich and Geneva. Switzerland's economic stability and political neutrality combine to make the banks among the safest in the world. For many years, the guarantee of banking secrecy also attracted foreign depositors. The governments of many nations expressed concern that some of their citizens had deposited money in Swiss bank accounts without paying taxes on the earnings to their own country. In 2009, the Swiss government adopted guidelines to ensure that the necessary taxes are paid. Switzerland's insurance corporations also attract a great deal of foreign business.

Since the early 1800's, large numbers of tourists have come to Switzerland. Today, ski resorts in the Alps attract many vacationers. Because most of the resorts are located above the timber line, the ski runs are free of trees. In the summer, people enjoy hiking and mountain climbing. Water sports on Switzerland's beautiful lakes are also popular tourist attractions.

MEETING PLACE OF NATIONS

Many international organizations have made their headquarters in Switzerland because the country provides an impartial meeting place. Switzerland's neutrality, however, prevents its participation in many of these organizations.

For example, Switzerland did not join the United Nations (UN) when it was founded in 1945, because of concerns that UN membership could require military action that would violate the Swiss neutrality policy. Switzerland finally became a member of the UN in 2002.

As a nonpartisan country, Switzerland has served as a forum where representatives of many nations could meet and talk freely. Many ambassadors from developing countries prefer to attend conferences in Switzerland rather than in other nations. Swiss diplomats often function as peacemakers between states whose political disagreements prevent normal, direct relations.

Swiss membership

The Swiss are willing to join organizations that do not endanger their neutral position. In 1960, Switzerland helped form the European Free Trade Association, an economic organization of European nations. And in 1963, Switzerland joined the Council of Europe, an organization that seeks to promote closer unity for human rights and social progress. However, Swiss voters in 2001 declined to join the European Union, an economic and political partnership to which most European nations belong.

The Red Cross, an international organization, was founded by Swiss philanthropist Jean Henri Dunant in 1863. The Red Cross was established to relieve human suffering in time of war. Today, the Red Cross tries to prevent pain and misery in peacetime as well. The organization's flag—a red cross on a white field—honors the Swiss flag, which features a white cross on a red field.

The International Red Cross is headquartered in Geneva, Switzerland. The flag for which the organization was named honors Switzerland's flag, a white cross on a red field. The Red Cross was founded in Switzerland.

The symbol of the Red Cross appears on a shack in Congo (Kinshasa). This Swiss-founded organization works to relieve suffering throughout the world.

Zurich—Switzerland's largest city—is home to many banks, including two of the country's largest: Credit Suisse Bank and USB. Swiss bankers are sometime referred as "the gnomes of Zurich" because like underground gnomes they conduct much of their business in secrecy.

Today, most nations have a Red Cross society. Most Muslim countries use a red crescent on a white field and call themselves Red Crescent societies. Israel uses a red Star of David. The national societies cooperate through the International Federation of Red Cross and Red Crescent Societies, headquartered in Geneva. The International Committee of the Red Cross, also located in Geneva, serves as a neutral intermediary during international conflicts. It also works for the continual improvement of the Geneva Conventions, which provide for humane treatment of wounded soldiers, prisoners of war, and civilian war victims.

Geneva headquarters

Many international organizations have their headquarters in Geneva. From 1920 until 1946, the Palace of Nations in Geneva served as headquarters for the League of Nations—an international peacekeeping organization with goals similar to those of the UN. Today, the palace houses the UN's European headquarters.

The International Labor Organization, the World Health Organization (WHO), and the World Council of Churches also have their headquarters in Geneva. And many international pacts, such as the Geneva Accords and the Geneva Conventions, were negotiated in the city.

The international organizations and firms based in Geneva provide many jobs for the city's residents and its guest workers. Guest workers from other nations make up about 30 percent of the city's population. Many people from foreign countries serve as officials of the international organizations headquartered in Geneva.

The research laboratory of CERN—the European Organization for Nuclear Research—stands just outside Geneva. Scientists from all over the world come to work on projects at this world-renowned facility.

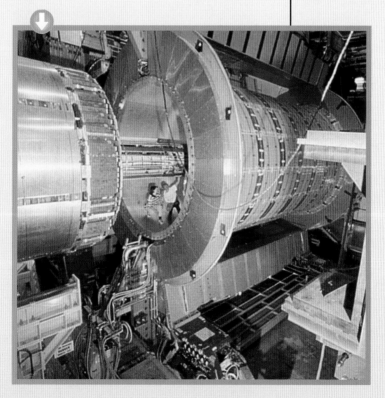

Syria is an Arab country on the eastern shore of the Mediterranean Sea. It lies south of Turkey and northwest of Iraq. Jordan is south of Syria, and Lebanon and Israel are to the southwest.

Since ancient times, Syria's plains, valleys, and deserts have been the sites of major trade routes between Asia and Mediterranean ports. Many empires and civilizations have risen and declined in Syria. Today, it is an independent nation with a rich cultural heritage.

Ancient cities

Syria forms the western portion of the Fertile Crescent, an arc-shaped region between the Mediterranean Sea, to the west, and the Persian Gulf, to the east. Throughout history, the region's rich soil has attracted many people. Some of the world's oldest civilizations developed in what is now Syria.

Archaeologists have found evidence of unidentified groups living in Syria before 4500 B.C. The first known settlers were Semites who probably arrived about 3500 B.C. They set up small farming villages along the fertile stretch of land between the mountains and grassy plains. In the 2000's B.C., the city-state of Ebla, with an economy based on agriculture and trade, became an important power in northern Syria.

Ancient Syria was a crossroads of trade. Camel caravans would set out from there for the long journey to China. The city of Palmyra, built in central Syria around a desert oasis, thrived more than 2,000 years ago as a major stop along the caravan route. Its ruins are a reminder of the great empire of Queen Zenobia. In A.D. 274, the Roman Emperor Aurelian defeated Queen Zenobia's army and destroyed the city.

Syria had become a province of the Roman Empire in 64 B.C. For nearly 700 years, Syrians lived under the Roman system of law, first as part of the Roman Empire, then as part of the Eastern Roman, or Byzantine, Empire. During this period, Christianity was born and developed in a part of Greater Syria called Palestine. Christianity became the state religion of Syria in the A.D. 300's.

SYRIA

Arabs and crusaders

In A.D. 636, invading Muslims from the Arabian Peninsula defeated the Byzantines. This invasion was to be a major turning point in the history of Syria.

Under the influence of the Muslims, the Latin and Greek languages of the Byzantines were replaced by Arabic. Christianity, the religion of the Byzantines, gave way to Islam. The city of Damascus served as the capital of a vast Islamic empire under the Umayyad dynasty, which lasted from 661 until the 700's. The religious and cultural traditions brought to Syria by the Arabs survive to this day.

Christian forces did not yield to Islam easily. In the late 1000's, Christian rulers in Europe launched a number of military expeditions to free Syria from Muslim rule and to regain the Holy Land (Palestine). These expeditions were called the Crusades. The crusaders were finally driven from Syria by Saladin, the Muslim ruler of Egypt, who then took control of most of Syria.

All that remains of the crusaders' efforts today are the ruins of their mighty fortresses along the Mediterranean coast. The giant Krak des Chevaliers, in the Jabal an Nusayriyah Mountains of Western Syria, is one of the few surviving relics of the "holy war."

The Mamluks, an Egyptian warrior dynasty from Egypt, ruled Syria from 1260 to 1516. The Mamluks were defeated by the Ottoman Empire, based in what is now Turkey.

The Ottomans ruled Syria for about 400 years, until their empire crumbled at the end of World War I (1914-1918). European powers then returned to Syrian shores. The League of Nations, an international association of countries formed in 1920, divided the region known as *Greater Syria* into four states—Syria, Lebanon, Palestine, and Transjordan. It gave control of Syria and Lebanon to France.

The Syrians did not like the presence of the French in their land, and they wanted to be an independent nation. But that was not to happen until the end of World War II (1939–1945). In 1946, France withdrew its troops, and the Syrians' dream of becoming an independent nation finally came true.

SYRIA TODAY

Since it became an independent nation in 1946, Syria has experienced much conflict both within its borders and with other countries in the Middle East. In 1947, the United Nations (UN) divided Palestine into a Jewish state—Israel—and an Arab state. That action began a long period of tension and fighting between Israel and the other Arab nations, including Syria.

Political developments

After Israel declared its independence in 1948, Syria and other Arab nations went to war to bring Israel's territory under Arab control. The Arab nations were unsuccessful, and the UN eventually arranged a cease-fire. Many Palestinians fled Israel and became refugees in neighboring Arab countries, including Syria.

Many Syrians blamed their government for failing to prevent the division of Palestine. In 1949, army officers overthrew the Syrian government. Over the next 20 years, a number of other military revolts upset the stability of Syria's government.

In 1958, Syria joined with Egypt in a political union called the United Arab Republic (U.A.R.). The group sought to promote Arab unity, but Syria withdrew in 1961 because of Egyptian domination of the union.

During the 1960's, the socialist Baath Party rose to power. In 1970, Hafez al-Assad, a Baathist air force commander, came to power in a military uprising. He was elected president the next year and remained in office until his death in 2000. His son, Bashar al-Assad, then became president. The Baath Party, with military backing, continued to control the Syrian government into the 2000's.

In 2011, antigovernment protests erupted in Syria. Protesters called for democratic reforms and the removal of President Assad. More than 5,000 people were killed in clashes between protesters and government forces. Despite pressure from the international community, the government continued its violent response to the protests.

FACTS

Official name:	Al Jumhuria Al-Arabia Al-Suria (Syrian Arab Republic)
Capital:	Damascus
Terrain:	Primarily semiarid and desert plateau; narrow coastal plain; mountains in west
Area:	71,498 mi² (185,180 km²)
Climate:	Mostly desert; hot, dry, sunny summers (June to August) and mild, rainy winters (December to February) along coast; cold weather with snow or sleet periodically hitting Damascus
Main rivers:	Euphrates, Khabur, Orontes, Balikh
Highest elevation:	Mount Hermon, 9,232 ft (2,814 m)
Lowest elevation:	Mediterranean Sea, sea level
Form of government:	Republic
Head of state:	President
Head of government:	Prime minister
Administrative areas:	13 muhafazat (provinces), Damascus
Legislature:	Majlis al-shaab (People's Council) with 250 members serving four-year terms
Court system:	Supreme Constitutional Court, Court of Cassation
Armed forces:	292,600 troops
National holiday:	Independence Day - April 17 (1946)
Estimated 2010 population:	21,399,000
Population density:	299 persons per mi² (116 per km²)
Population distribution:	51% urban, 49% rural
Life expectancy in years:	Male, 70; female, 74
Doctors per 1,000 people:	0.5
Birth rate per 1,000:	27
Death rate per 1,000:	4
Infant mortality:	19 deaths per 1,000 live births
Age structure:	0-14: 36%; 15-64: 61%; 65 and over: 3%
Internet users per 100 people:	17
Internet code:	.sy
Languages spoken:	Arabic (official), Kurdish, Armenian, Aramaic, Circassian, French, English
Religions:	Sunni Muslim 74%, other Muslim 16%, Christian 10%, Jewish
Currency:	Syrian pound
Gross domestic product (GDP) in 2008:	$54.80 billion U.S.
Real annual growth rate (2008):	4.8%
GDP per capita (2008):	$2,683 U.S.
Goods exported:	Clothing, cotton, crude oil, food, textiles
Goods imported:	Chemicals, food, machinery, metal and metal products, petroleum products, transportation equipment
Trading partners:	China, France, Italy, Lebanon, Saudi Arabia, Turkey

The republic of Syria lies along the eastern Mediterranean coast. Mountains divide the coastal region from the country's interior, which is covered by deserts and dry grasslands. Once part of the Ottoman Empire and later controlled by France, Syria gained its independence in 1946.

TURKEY

Dayrik
Amuda Al Qamishli
Ayn al Arab
Jarabulus
Tall Abyad Ras al Ayn
Azaz Jawban Bayk TALL HALAF
Tall Akhtarin Manbij 3,018 ft (920 m) Al Hasakah
Afrin Rifat Al Bab JABAL ABD AL AZIZ
Muslimiyah Ayn Isa Shaddadi
Dar Taizzah Aleppo As Lake Assad
Harim Safirah Ar Raqqah Fadghami
Salqin Maskanah
Idlib Taftanaz Khanasir As Sabkhah
Kassab Saraqib Madinat ath Thawrah TABKA DAM
Jis ash Shughur Ariha AR RUSAFAH
Latakia EBLA Khan Shaykhun As, Suwar
Jablah APAMEA Maarrat an Numan JABAL BISHRI
Baniyas Muhradah Dayr az Zawr
Tartus Masyaf Hama Al Mayadin
Burj Safita As Salamiyah 4,560 ft (1,390 m) As Sukhnah As Salihiyah
KRAK DES CHEVALIERS Ar Rastan JABAL ABU RUJMAYN MARKO
Homs Furqlus As Salihiyah Abu Kamal
Tall Kalakh Al Qusayr Tiyas Tudmur PALMYRA
Mediterranean Sea Al Burayj Sadad PALMYRA
LEBANON Dayr Atiyah IRAQ
An Nabk JABAL AR RUWAQ
Sirghaya Yabrud Sab Biar
Az Zabdani Jayrud
Al Qutayfah
Jawban Dumayr Khan Abu Shamat North
Mt. Hermon Darayya Duma
9,232 ft (2,814 m) Damascus
Baniyas Mukhayyam al Yarmuk SYRIAN DESERT
Al Qunaytirah 0 50 100 Miles
Golan Heights Nawa 0 50 100 Kilometers
Izra Shahba
As Jabal ad Duruz
Dara Suwayda 5,909 ft (1,801 m)
Busra ash Sham
Salkhad

CYPRUS

ISRAEL

West Bank

Gaza Strip

Dead Sea

JORDAN

SAUDI ARABIA

In the Syrian city of Aleppo, the busy Al-Gassaniyeh street swarms with traders and business people. Aleppo ranks among Syria's largest cities. It has been an important trading center since the 1500's B.C.

Regional conflicts

In June 1967, Israel went to war with Syria, Jordan, and Egypt. After six days of fighting, Israel had won the war and occupied much Arab land, including the Golan Heights in the far southwestern corner of Syria. Another brief war in 1973 failed to regain the territory for Syria. Israel claimed legal and political authority in the Golan Heights in 1981, but Syria and other nations denounced the action. Clashes between Israel and Syria have continued in the region.

In 1976, Syria sent troops into Lebanon to stop a civil war there. The civil war ended in 1991. However, the Syrian forces remained. In 2005, Syria withdrew the troops under pressure from groups in Lebanon and other countries.

In August 1990, Iraq invaded and occupied Kuwait. After the invasion, Syria, the United States, and many other countries formed an alliance to oppose Iraq's occupation of Kuwait. War broke out in January 1991, and the allies defeated Iraq in February. About 20,000 Syrian troops took part in the Persian Gulf War of 1991.

LAND AND ECONOMY

Syria is a land of rolling plains, fertile river valleys, and barren deserts. The country is located at the western end of a rich farmland called the Fertile Crescent.

Landscape

Syria's landscape falls into three main categories: (1) the coast, (2) the mountains, and (3) the valleys and plains.

The coast of Syria is a narrow strip of land along the eastern shore of the Mediterranean Sea. It stretches from Turkey in the north to Lebanon in the south. The land and climate along the coast are typical of the Mediterranean region.

Moist sea winds give the region mild, humid weather conditions. Temperatures average about 48° F (9° C) in January and about 81° F (27° C) in July. About 40 inches (100 centimeters) of rain falls yearly. Because of the humid conditions, the coast is one of the few areas of Syria that does not need irrigation to grow crops.

The mountains run mostly from north to south. The Jabal an Nusayriyah range rises east of the coast. The Anti-Lebanon range extends along the border with Lebanon. It contains Mount Hermon, the highest peak in Syria. Southeast of the Anti-Lebanon range is Jabal ad Duruz, a mountain in the plateau region at the southeastern corner of the country.

The mountains catch the sea winds blowing inland from the Mediterranean and force them to drop their moisture on the western side. The western side of the Jabal an Nusayriyah and of Jabal ad Duruz receive abundant rainfall, and the soil is rich and fertile. The regions are well populated with farmers. On the eastern side, though, the air is dry, and the land has a stony surface. The Anti-Lebanon Mountains, which lie to the east of another mountain range in Lebanon, are dry and less populated.

Farther east lie the valleys and plains, the third and largest region of Syria. This region includes fertile river valleys, grassy plains, and sandy deserts.

The water wheel on the Orontes River is part of an ancient irrigation system. Because Syria receives so little rainfall, farmers must depend on irrigation to water their crops.

A Syrian farmer harvests a field of squash in northwest Syria, where the waters of the Euphrates River irrigate the fields.

The Orontes River and mountain streams bring water to the rich farmland along the mountain slopes. In northeast Syria, the Euphrates River and its tributaries provide water to the developing agricultural area in that region. In addition, the Tabka Dam on the Euphrates provides hydroelectric power. Most of the rest of Syria is covered by deserts and dry grasslands.

Agriculture and economy

Agricultural land and petroleum are Syria's two most valuable natural resources. Agriculture accounts for about 20 percent of the total value of Syria's economic production. Petroleum is Syria's chief mineral product, and mining makes up about 7 percent of Syria's production value.

Syrian farmers depend on irrigation to provide the necessary moisture for their crops because rainfall is too light and irregular. Cotton and wheat are Syria's main crops. Farmers also grow barley, sugar beets, and tobacco. Their fruit and vegetable crops include grapes, olives, and tomatoes. A few nomadic Bedouins raise herds of goats and sheep.

Oil provides more than half of Syria's export earnings and much of its energy needs. The country's other mineral products include gypsum, limestone, and natural gas.

One of Syria's chief industries is the manufacture of cotton fabrics and other textiles. Other important products include beverages, cement, fertilizer, glass, processed foods, and sugar. Syria has a growing oil refining industry.

Service industries account for about 60 percent of the value of Syria's economic production and employ about 40 percent of its workers. The leading service industries are wholesale and retail trade, tourism, and government services. Aleppo, Damascus, and Latakia are the leading trade centers.

The government controls most of the economy, but the majority of farms, small businesses, and small industries are privately owned. In the 1990's, the government adopted a policy of reducing economic restrictions and allowing more privatization.

The land of northern Syria is mostly cultivated by farmers who work small plots. Government funds have helped many farmers replace their old-fashioned equipment with modern machinery.

A quiet courtyard in Aleppo's market quarter is part of a caravanserai, a rest stop for caravans on their routes between Asia and the Mediterranean. Caravans of as many as 5,000 camels once brought goods out of the Far East across the Syrian plains to seaports for shipment to Europe.

PEOPLE

Syria has more than 21 million people. About 90 percent of them speak Arabic, the country's official language, and consider themselves Arabs. About 90 percent of the people also are Muslims. However, even though the overwhelming majority of people follow Islamic tradition, Syria still contains a number of other religious and ethnic groups. As a result, the country has an interesting mixture of cultural traditions.

The Syrian people trace their origins to the Semitic people who lived in ancient Syria more than 5,000 years ago. During the following centuries, many groups—including Hebrews, Assyrians, Babylonians, Greeks, Romans, Arabs, and Egyptians—all left their mark on the country. Each group has made a contribution to Syria's long and colorful past.

Ethnic minorities

Non-Arab Syrians include Armenians and Kurds. Their ancestors came from the north. Most of these Syrians still speak Armenian or Kurdish in everyday life.

The Kurds live mainly in rural villages, farming the land and raising sheep and goats. Their homeland extends beyond Syria over the mountainous region of Southwest Asia, but they have never had their own separate state and government. The Kurds would like to be culturally and politically independent, and they have often clashed with the governments of the countries in which they live. The majority of Kurds are Sunni Muslims. The Kurdish language is an Indo-European language closely related to Persian.

The ancestors of the Armenians were a highly cultured ancient people who had converted to Christianity by the A.D. 300's. Today the Armenians live mainly in Aleppo.

Religious life

Most Muslims in Syria belong to the Sunni branch of Islam. Members of some smaller Muslim groups, including Alawites and Shiites, also live in Syria. Some Syrians are Druses, who practice a secret religion related to Islam. About 10 percent of the population is Christian, and there are a small number of Syrian Jews.

Religion, especially Islam, is a powerful political and social force in Syria.

In the remote mountain village of Malula in northern Syria, members of the country's Christian minority still speak the ancient Semitic language of Aramaic, also called Assyrian. Jesus Christ and his earliest followers spoke Aramaic.

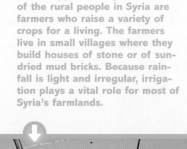

Syrian farmers prepare a pump near Aleppo for irrigation. Most of the rural people in Syria are farmers who raise a variety of crops for a living. The farmers live in small villages where they build houses of stone or of sun-dried mud bricks. Because rainfall is light and irregular, irrigation plays a vital role for most of Syria's farmlands.

Much of Syria's rural population is concentrated in northern villages like Halawa. People in such rural areas often wear more traditional clothing.

The historic Bakdash ice cream shop serves customers in the Souq al-Hamidiyya, a long, covered market that leads into the heart of the Old City of Damascus. The shop, which opened in 1895, serves thousands of people daily.

However, the ruling Baath Party is not officially tied to any religion, and some Christians serve in leading government positions. Michel Aflaq, who developed the principles of Arab socialism and the Baath Party, was a Christian.

Syria's wealth of religious and ethnic variety has given the country a rich and diverse culture. Yet tension between these groups has also threatened Syria's peace and stability.

Way of life

Syria has some of the world's oldest cities. The ancient city centers have narrow, winding streets and ancient marketplaces. The new sections of Syria's cities, however, resemble modern cities around the world, and many people live in apartment buildings. Most rural people live in houses of stone or sun-dried mud bricks in small villages.

Family ties are close among most Syrians. Many parents share their home with their sons and the sons' families. It is becoming increasingly common, however, for married children and their families to live apart from grandparents, aunts, and uncles.

Women in Syria have faced great challenges in gaining social privileges and freedoms. However, increasing educational opportunities and new attitudes are improving the position of women. A number of Syrian women now occupy prominent roles in government, Baath Party leadership, and society in general.

DAMASCUS

Damascus is the capital of Syria, as well as its cultural, political, and economic center. Historians believe that Damascus may have been founded about 5,000 years ago, making it one of the world's oldest cities.

The city is built on an oasis in a semiarid plain. The Barada River, which flows through Damascus, has provided the city's water for thousands of years.

Damascus is a city of contrasts. Its southern section includes an area of narrow, winding streets that are hundreds of years old. In *suqs* (bazaars) along these streets, merchants sell their wares to passers-by in the age-old tradition. The business district in the northwestern section of the city has a more modern look. It features tall buildings erected since the 1900's.

A jewel of Islam

Damascus was an important city during the rule of several ancient empires. The Assyrians, Greeks, Romans, and Byzantines all played a role in the history of Damascus.

The city was part of the Christian Byzantine Empire when the Muslim Arab armies of Khalid ibn al-Walid invaded it in A.D. 635. The Byzantine Emperor Heraclius had been a very unpopular ruler. So, when Damascus fell to the Muslims, its people willingly accepted the conquerors.

The Muslim Arabs made Damascus the capital of the Umayyad dynasty in A.D. 661. The city thrived under Umayyad leadership until the dynasty lost control of it in the 700's.

Saladin, the Muslim ruler of Egypt who drove the Christian crusaders out of Syria, took control of Damascus in the late 1100's. In 1516, the Ottoman Empire conquered Damascus. This powerful empire controlled the region until World War I, when the combined forces of Syrians, Arabs, and the United Kingdom captured Damascus in 1918. France acquired control of Syria in 1920.

The Moorish Gardens of Damascus show the elegance of Damascus in all its ancient glory. The city has been an important trade center for centuries. Many empires and dynasties have ruled Damascus since it was founded about 3000 B.C. Each culture has left its influence on the city.

In 1946, Syria became an independent nation, with Damascus as its capital. Over the decades that followed, the city's population grew dramatically. The government established new towns nearby to help with the resulting housing shortage.

Historic sites

Although much of old Damascus has been cleared away to make room for a swelling population, traces of its former glory can still be seen around the city.

The Azem Palace, built in the mid-1700's, stands in the middle of the old city. The palace had separate buildings for men and women, each with a series of bathing rooms. The palace now houses a museum. Beyond the palace, the al-Hamadieh suq beckons to visitors with its wide variety of hand-crafted goods. In crowded booths, craftworkers sell objects made of brass, wood, straw, leather, silver, and gold.

The Umayyad Mosque, also called the Great Mosque, stands at the end of the arches of the al-Hamadieh suq. The Great Mosque was built in the early 700's by Caliph al-Walid I. It was built on the site of the Christian Church of St. John the Baptist, which had, in turn, been built on the site of a Roman temple to Jupiter. The mosque's mosaics and statues are a mix of Byzantine and Muslim Umayyad styles.

Located near the al-Hamadieh suq is the Takiyya as-Süleimaniyya, built in the 1500's during the Ottoman period. Today, this complex of buildings includes a mosque, a museum, and a handicrafts bazaar. Damascus also features museums, a university, and several theaters.

With its historic sites and cultural activities, Damascus is a popular tourist destination. However, economic and political troubles have slowed government efforts to further develop Syria's tourism industry.

The Umayyad Mosque stands on the ruins of the Christian Church of St. John the Baptist. It was built in the A.D. 700's and later restored.

The al-Hamadieh suq buzzes with life and energy. A covered bazaar, it offers a wide range of hand-crafted items by local workers. The suq also serves as a social center, where local merchants and business people gather to discuss the business of the day.

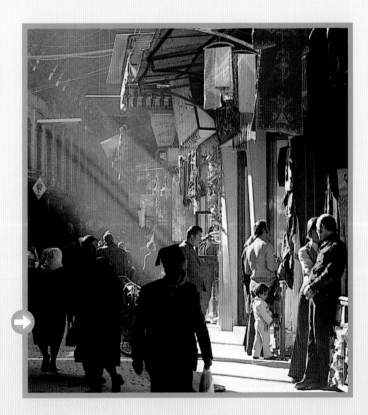